Practical Guide to **Logistic Regression**

Practical Guide to Logistic Regression

Joseph M. Hilbe

Jet Propulsion Laboratory
California Institute of Technology, USA

and

Arizona State University, USA

CRC Press is an imprint of the
Taylor & Francis Group, an **informa** business
A CHAPMAN & HALL BOOK

CRC Press
Taylor & Francis Group
6000 Broken Sound Parkway NW, Suite 300
Boca Raton, FL 33487-2742

Printed on acid-free paper
Version Date: 20150427

International Standard Book Number-13: 978-1-4987-0957-6 (Paperback)

Library of Congress Cataloging-in-Publication Data

Hilbe, Joseph M., 1944-
Practical guide to logistic regression / Joseph M. Hilbe.
pages cm
"A CRC title."
Includes bibliographical references and index.
ISBN 978-1-4987-0957-6 (pbk. : alk. paper) 1. Logistic regression analysis. 2. Regression analysis. 3. Multivariate analysis. 4. Statistics. I. Title.

QA278.2.H534 2016
519.5'36--dc23
2015012864

Visit the Taylor & Francis Web site at
http://www.taylorandfrancis.com

and the CRC Press Web site at
http://www.crcpress.com

Contents

Preface ix
Author xv

1 Statistical Models **1**
1.1 What Is a Statistical Model? 1
1.2 Basics of Logistic Regression Modeling 3
1.3 The Bernoulli Distribution 4
1.4 Methods of Estimation 7
SAS Code 11
Stata Code 12

2 Logistic Models: Single Predictor **13**
2.1 Models with a Binary Predictor 13
2.2 Predictions, Probabilities, and Odds Ratios 18
2.3 Basic Model Statistics 20
2.3.1 Standard Errors 20
2.3.2 *z* Statistics 23
2.3.3 *p*-Values 23
2.3.4 Confidence Intervals 24
2.4 Models with a Categorical Predictor 28
2.5 Models with a Continuous Predictor 32
2.5.1 Varieties of Continuous Predictors 32
2.5.2 A Simple GAM 33
2.5.3 Centering 34
2.5.4 Standardization 36
2.6 Prediction 37
2.6.1 Basics of Model Prediction 37
2.6.2 Prediction Confidence Intervals 39
SAS Code 41
Stata Code 47

3 Logistic Models: Multiple Predictors **49**
3.1 Selection and Interpretation of Predictors 49
3.2 Statistics in a Logistic Model 52

3.3 Information Criterion Tests 58
3.3.1 Akaike Information Criterion 58
3.3.2 Finite Sample 59
3.3.3 Bayesian Information Criterion 60
3.3.4 Other Information Criterion Tests 60
3.4 The Model Fitting Process: Adjusting Standard Errors 61
3.4.1 Scaling Standard Errors 61
3.4.2 Robust or Sandwich Variance Estimators 63
3.4.3 Bootstrapping 64
3.5 Risk Factors, Confounders, Effect Modifiers, and Interactions 65
SAS Code 67
Stata Code 70

4 Testing and Fitting a Logistic Model 71
4.1 Checking Logistic Model Fit 71
4.1.1 Pearson *Chi*2 Goodness-of-Fit Test 71
4.1.2 Likelihood Ratio Test 72
4.1.3 Residual Analysis 73
4.1.4 Conditional Effects Plot 79
4.2 Classification Statistics 81
4.2.1 S–S Plot 84
4.2.2 ROC Analysis 84
4.2.3 Confusion Matrix 86
4.3 Hosmer–Lemeshow Statistic 88
4.4 Models with Unbalanced Data and Perfect Prediction 91
4.5 Exact Logistic Regression 93
4.6 Modeling Table Data 96
SAS Code 101
Stata Code 105

5 Grouped Logistic Regression 107
5.1 The Binomial Probability Distribution Function 107
5.2 From Observation to Grouped Data 109
5.3 Identifying and Adjusting for Extra Dispersion 113
5.4 Modeling and Interpretation of Grouped Logistic Regression 115
5.5 Beta-Binomial Regression 117
SAS Code 123
Stata Code 125

6 Bayesian Logistic Regression **127**
6.1 A Brief Overview of Bayesian Methodology 127
6.2 Examples: Bayesian Logistic Regression 130
6.2.1 Bayesian Logistic Regression Using R 130
6.2.2 Bayesian Logistic Regression Using JAGS 137
6.2.3 Bayesian Logistic Regression with Informative Priors 143
SAS Code 147
Stata Code 148
Concluding Comments 149

References 151
Index 153

Preface

Logistic regression is one of the most used statistical procedures in research. It is a component of nearly all, if not all, general purpose commercial statistical packages, and is regarded as one of the most important statistical routines in fields such as health-care analysis, medical statistics, credit rating, ecology, social statistics and econometrics, and other similar areas. Logistic regression has also been considered by many analysts to be an important procedure in predictive analytics, as well as in the longer established Sigma Six movement.

There is a good reason for this popularity. Unlike traditional linear or normal regression, logistic regression is appropriate for modeling a binary variable. As we shall discuss in more detail in the first chapter, a binary variable has only two values—1 and 0. These values may be thought of as "success" and "failure," or of any other type of "positive" and "non-positive" dichotomy. If an analyst models a 1/0 binary variable on one or more predictors using linear regression, the assumptions upon which the linear model is based are violated. That is, the linear model taught in Introduction to Statistics courses is not appropriate for modeling binary data. We shall discuss why this is the case later in the book.

Logistic regression is typically used by researchers and analysts in general for three purposes:

1. To *predict* the probability that the outcome or response variable equals 1
2. To *categorize* outcomes or predictions
3. To *access* the odds or risk associated with model predictors

The logistic model is unique in being able to accommodate all three of these goals. The foremost emphasis of this book is to help guide the analyst in utilizing the capabilities of the logistic model, and thereby to help analysts to better understand their data, to make appropriate predictions and classifications, and to determine the odds of one value of a predictor compared to another. In addition, I shall recommend an approach to logistic regression modeling that satisfies problems that some "data science" analysts find with traditional logistic modeling.

This book is aimed at the working analyst or researcher who finds that they need some guidance when modeling binary response data. It is also of value for those who have not used logistic regression in the past, and who are not familiar with how it is to be implemented. I assume, however, that the reader has taken a basic course in statistics, including instruction on applying linear regression to study data. It is sufficient if you have learned this on your own. There are a number of excellent books and free online tutorials related to regression that can provide this background.

I think of this book as a basic guidebook, as well as a tutorial between you and me. I have spent many years teaching logistic regression, using logistic-based models in research, and writing books and articles about the subject. I have applied logistic regression in a wide variety of contexts—for medical and health outcomes research, in ecology, fisheries, astronomy, transportation, insurance, economics, recreation, sports, and in a number of other areas. Since 2003, I have also taught both the month-long Logistic Regression and Advanced Logistic Regression courses for *Statistics.com*, a comprehensive online statistical education program. Throughout this process I have learned what the stumbling blocks and problem areas are for most analysts when using logistic regression to model data. Since those taking my courses are located at research sites and universities throughout the world, I have been able to gain a rather synoptic view of the methodology and of its use in research in a wide variety of applications.

In this volume, I share with you my experiences in using logistic regression, and aim to provide you with the fundamental logic of the model and its appropriate application. I have written it to be the book I wish I had read when first learning about the model. It is much smaller and concise than my 656 page *Logistic Regression Models* (Chapman & Hall/CRC, 2009), which is a general reference to the full range of logistic-based models. Rather, this book focuses on how best to *understand* the key points of the basic logistic regression model and how to use it properly to model a binary response variable. I do not discuss the esoteric details of estimation or provide detailed analysis of the literature regarding various modeling strategies in this volume, but rather I focus on the most important features of the logistic model—how to construct a logistic model, how to interpret coefficients and odds ratios, how to predict probabilities based on the model, and how to evaluate the model as to its fit. I also provide a final chapter on Bayesian logistic regression, providing an overview of how it differs from the traditional frequentist tradition. An important component of our examination of Bayesian modeling will be a step-by-step guide through JAGS code for modeling real German health outcomes data. The reader should be able to attain a basic understanding of how Bayesian logistic regression models can be developed and interpreted—and be able to develop their own models using the explanation in the book as a guideline.

Resources for how to learn how to model slightly more complicated models will be provided—where to go for the next step. Bayesian modeling is having a continually increasing role in research, and every analyst should at least become acquainted with how to understand this class of models, and with how to program basic Bayesian logistic models when doing so is advisable.

R statistical software is used to display all but one statistical model discussed in the book—exact logistic regression. Otherwise R is used for all data management, models, postestimation fit analyses, tests, and graphics related to our discussion of logistic regression in the book. SAS and Stata code for all examples is provided at the conclusion of each chapter. Complete Stata and SAS code and output, including graphics and tables, is provided on the book's web site. R code is also provided on the book's web site, as well as in the LOGIT package posted on CRAN.

R is used in the majority of newly published texts on statistics, as well as for examples in most articles found in statistics journals published since 2005. R is open ware, meaning that it is possible for users to inspect the actual code used in the analysis and modeling process. It is also free, costing nothing to download into one's computer. A host of free resources is available to learn R, and blogs exist that can be used to ask others how to perform various operations. It is currently the most popular statistical software worldwide; hence, it makes sense to use it for examples in this relatively brief monograph on logistic regression. But as indicated, SAS and Stata users have the complete code to replicate all of the R examples in the text itself. The code is in both printed format as well as electronic format for immediate download and use.

A caveat: Keep in mind that when copying code from a PDF document, or even from a document using a different font from that which is compatible with R or Stata, you will likely find that a few characters need to be retyped in order to successfully execute. For example, when pasting program code from a PDF or word document into the R editor, characters such as "quotation marks" and "minus signs" may not convert properly. To remedy this, you need to retype the quotation or minus sign in the code you are using.

It is also important to remember that this monograph is not about R, or any specific statistical software package. We will foremost be interested in the logic of logistic modeling. The examples displayed are aimed to clarify the modeling process. The R language, although popular and powerful, is nevertheless tricky. It is easy to make mistakes, and R is rather unforgiving when you do. I therefore give some space to explaining the R code used in the modeling and evaluative process when the code may not be clear. The goal is to provide you with code you can use directly, or adapt as needed, in order to make your modeling tasks both easier and more productive.

I have chosen to provide Stata code at the end of each chapter since Stata is one of the most popular and to my mind powerful statistical packages on the commercial market. It has free technical support and well-used blog and user LISTSERV sites. In addition, it is relatively easy to program statistical procedures and tests yourself using Stata's programming language. As a result, Stata has more programs devoted to varieties of logistic-based routines than any other statistical package. Bob Muenchen of the University of Tennessee and I have pointed out similarities and differences between Stata and R in our 530 page book, *R for Stata Users* (Springer, 2010). It is a book to help Stata users learn R, and for R users to more easily learn Stata. The book is published in hardback, paperback, and electronic formats.

I should acknowledge that I have used Stata for over a quarter of a century, authoring the initial versions of several procedures now in commercial Stata including the first *logistic* (1990) and *glm* (1993) commands. I also founded the *Stata Technical Bulletin* in 1991, serving as its first editor. The STB became enhanced to the *Stata Journal* in 1999. I also used to teach S-Plus courses for the manufacturer of the package in the late 1980s and early 1990s, traveling to various sites in the United States and Canada for some 4 years. The S and S-Plus communities have largely evolved to become R users during the past decade to decade and a half. In addition, I also programmed various macros in SAS and gave presentations at SUGI, thus have a background in SAS as well. However, since it has been a while since I have used SAS on a regular basis, I invited Yang Liu, a professional SAS programmer and MS statistician to replicate the R code used for examples in the text into SAS. He has provided the reader with complete programming code, not just snippets of code that one finds in many other texts. The SAS/Stat GENMOD Procedure and Proc Logistic were the two most used SAS procedures for this project. Yang also reviewed proof pages with me, checking for needed amendments.

The R data sets and user authored functions and scripts are available for download and installation from the CRAN package, LOGIT. The LOGIT package will also have the data, functions, and scripts for both the first (2009) and second (forthcoming 2016) edition of the author's *Logistic Regression Models* (Chapman & Hall/CRC). Data files in Stata, SAS, SPSS, Excel and csv format, as well as Stata commands and ado/do files are located on the author's web site:

http://works.bepress.com/joseph_hilbe/
as well as on the publishers web site for the book:
http://www.crcpress.com/product/isbn/9781498709576

An *Errata and Comments* PDF as well as other resource material and "handouts" related to logistic regression will also be available on my *Bepress* web site.

I wish to acknowledge the following colleagues for their input into the creation of this book. Rafael S. de Souza (astrophysicist, Eötvös Loránd University, Hungary) and Yang Liu (Baylor Scott & White Health). My collaborative work with James W. Hardin (associate professor of Biostatistics, University of South Carolina) over the past 15 years has indirectly contributed to this book as well. Our collaboration has involved coauthoring five books, a number of book chapters and journal articles, and numerous discussions on statistical programming and modeling. My work with Alain Zuur (CEO, Highlands Statistics, Scotland, UK) also contributed to this book. We coauthored a book in 2013 related to Bayesian modeling that has greatly influenced my subsequent work in the area. I should also acknowledge Peter Bruce (CEO, Institute for Statistics Education, *Statistics.com*), who first suggested that I write this book for use in my web course on Logistic Regression. Dr. de Souza provided two new R functions he authored for the book's classification statistics and graphics in Chapter 4 called *ROCtest* and *confusion_stat*. These are very nicely written and useful functions that enhance R's logistic modeling capabilities. Yang Liu is responsible for all of the SAS code provided in the book, testing it against the R functions, tests, and graphics presented throughout the text. He also wrote SAS code and full output for all examples in the text, which are on the book's website, and thoroughly reviewed the entire book for errata and suggested amendments at the proof stage. I also acknowledge Shawn Casper (Managing Director, Praxis Reliability Consulting, LLC, Monroe, MI) who also read the entire manuscript, checking text and code, and offering a number of helpful suggestions to the book, and Dr. Jamie Riggs (Predictive Analytics Masters Program, Northwestern University) for reviewing early chapters when the book started to take form. I need to also mention Judith M. Simon, Project Editor, CRC Press, who was responsible for the overall production of the book, and Syed Mohamad Shajahan, Deputy Manager, Techset Composition, Chennai, India, who was responsible for the actual page set-up and production of the book. Both did an outstanding job in helping create this book, and in tolerating the many amendments I made to it. Robert Calver, statistics editor at Chapman & Hall/CRC, has been more helpful than I can express here. He has been my editor since 2002, a position at which he truly excels, and is a good friend.

I dedicate this text to Heidi and Sirr Hilbe. Heidi died over 40 years ago, but was my best companion at the time I authored my first text in 1970 some 45 years ago, and warrants my recognition. Sirr has been my constant companion since his birth in 2007, and keeps me sane as I write yet another book. Sirr is a small white Maltese, but this takes nothing away from his unique contribution to this text.

Joseph M. Hilbe
Florence, AZ

Author

Joseph M. Hilbe (1944–) is a Solar System Ambassador with NASA's Jet Propulsion Laboratory, California Institute of Technology, an adjunct professor of statistics at Arizona State University, and emeritus professor at the University of Hawaii. He is currently president of the International Astrostatistics Association, is an elected Fellow of the American Statistical Association, is an Elected Member of the International Statistical Institute and Full Member of the American Astronomical Society. Professor Hilbe is one of the leading statisticians in modeling discrete and longitudinal data, and has authored a number of books in these areas including best sellers, *Logistic Regression Models* (Chapman & Hall/CRC, 2009), two editions of *Negative Binomial Regression* (Cambridge University Press, 2007, 2011), and *Modeling Count Data* (Cambridge University Press, 2014).

Other statistics books by Joseph M. Hilbe:

- *Solutions Manual for Logistic Regression Models* (2009)
- *R for Stata Users* (2010; with R. Muenchen)
- *A Beginner's Guide to GLM and GLMM with R: A Frequentist and Bayesian Perspective for Ecologists* (2013; with A. Zuur and E. Ieno)
- *Astrostatistical Challenges for the New Astronomy* (2013)
- *Methods of Statistical Model Estimation* (2013; with A. Robinson)
- *Generalized Estimating Equations* (2003, 2013; with J. Hardin)
- *Quasi-Least Squares Regression* (2014; with J. Shults)
- *Practical Predictive Analytics and Decisioning Systems for Medicine* (2015; with L. Miner, P. Bolding, M. Goldstein, T. Hill, R. Nesbit, N. Walton, and G. Miner)
- *Generalized Linear Models and Extensions* (2001, 2007, 2013, 2015; with J. Hardin)

Statistical Models

1

> **Statistics**: *Statistics may generically be understood as the science of collecting and analyzing data for the purpose of classification, prediction, and of attempting to quantify and understand the uncertainty inherent in phenomena underlying data*
> (Hilbe, 2014)

1.1 WHAT IS A STATISTICAL MODEL?

A model is typically thought of as a simplification of a more complex situation. The focus of a model is to abstract the most important or key features from what is being modeled so that we may more clearly understand the modeled situation, or see how it relates to other aspects of reality. There are a variety of different types of models though, but each type still represents an approximation or simplification of something more detailed.

Statistics deals with data, which can be notoriously messy and complex. A statistical model is a simplification of some data situation, whether the data are about the weather, health outcomes, or the number of frogs killed on a highway over a period of a year. Data can be about nearly anything that can be measured or tested. In order to be measured though, data must be numerically expressed; that is, a statistical model is a means to simplify or clarify numbers.

The models we are going to be discussing in this monograph are called parametric statistical models. As such they are each based on an underlying probability distribution. Since probability distributions are characterized and defined by parameters, models based on them are referred to as parametric models. The fundamental idea of a parametric model is that the data to be modeled by an analyst are in fact generated by an underlying probability distribution function or PDF. The analyst does not usually observe the entire range of data defined by the underlying PDF, called the population data, but rather observes a random sample from the underlying data. If the sample of data is truly representative of the population data, the sample data will be described by the same PDF as the population data, and have the same values of its parameters, which are initially unknown.

Parameters define the specific mean or location (shape) and perhaps scale of the PDF that best describes the population data, as well as the distribution of the random sample from the population. A statistical model is the relationship between the parameters of the underlying PDF of the population data and the estimates made by an analyst of those parameters.

Regression is one of the most common ways of estimating the true parameters in as unbiased manner as possible. That is, regression is typically used to establish an accurate model of the population data. Measurement error can creep into the calculations at nearly every step, and the random sample we are testing may not fully resemble the underlying population of data, nor its true parameters. The regression modeling process is a method used to understand and control the uncertainty inherent in estimating the true parameters of the distribution describing the population data. This is important since the predictions we make from a model are assumed to come from this population.

Finally, there are typically only a limited range of PDFs which analysts use to describe the population data, from which the data we are analyzing is assumed to be derived. If the variable we are modeling, called the response term (y), is binary (0,1), then we will want to use a Bernoulli probability distribution to describe the data. The Bernoulli distribution, as we discuss in more detail in the next section, consists of a series of 1s and 0s. If the variable we wish to model is continuous and appears normally distributed, then we assume that it can be best modeled using a Gaussian (normal) distribution. This is a pretty straightforward relationship. Other probability distributions commonly used in modeling are the lognormal, binomial, exponential, Poisson, negative binomial, gamma, inverse Gaussian, and beta PDFs. Mixtures of distributions are also constructed to describe data. The lognormal, negative binomial, and beta binomial distributions are such mixture distributions—but they are nevertheless completely valid PDFs and have the same basic assumptions as do other PDFs.

I should also mention that probability distributions do not all have the same parameters. The Bernoulli, exponential, and Poisson distributions are single parameter distributions, and models directly based on them are single parameter models. That parameter is the mean or location parameter. The normal, lognormal, gamma, inverse Gaussian, beta, beta binomial, binomial, and negative binomial distributions are two parameter models. The first four of these are continuous distributions with mean (shape) and scale (variability) parameters. The binomial, beta, and beta binomial distributions will be discussed later when discussing grouped logistic regression.

The catcher in this is that a probability distribution has various assumptions. If these assumptions are violated, the estimates we make of the parameters are biased, and may be incorrect. Statisticians have worked out a number of adjustments for what may be called "violations of distributional assumptions," which are important for an analyst to use when modeling data exhibiting problems. I'll mention these assumptions shortly, and we will address them in more detail as we progress through the book.

I fully realize that the above description of a statistical model—of a parametric statistical model—is not the way we normally understand the modeling process, and it may be a bit confusing. But it is in general the way statisticians think of statistical modeling, and is the basis of the frequency-based tradition of statistical modeling. Keep these relationships in mind as we describe logistic regression.

1.2 BASICS OF LOGISTIC REGRESSION MODELING

Logistic regression is foremost used to model a binary (0,1) variable based on one or more other variables, called predictors. The binary variable being modeled is generally referred to as the response variable, or the dependent variable. I shall use the term "response" for the variable being modeled since it has now become the preferred way of designating it. For a model to fit the data well, it is assumed that

The predictors are uncorrelated with one another.
That they are significantly related to the response.
That the observations or data elements of a model are also uncorrelated.

As discussed in the previous section, the response is also assumed to fit closely to an underlying probability distribution from which the response is a theoretical sample. The goal of a model is to estimate the true parameter(s) of the underlying PDF of the model based on the response as adjusted by its predictors. In the case of logistic regression, the response is binary (0,1) and follows a Bernoulli probability distribution. Since the Bernoulli distribution is a subset of the more general binomial distribution, logistic regression is recognized as a member of the binomial family of regression models. A comprehensive analysis of these relationships is provided in Hilbe (2009).

In this monograph, I assume that the reader is familiar with the basics of regression. However, I shall address the fundamentals of constructing, interpreting, fitting, and evaluating a logistic model in subsequent chapters. I shall also describe how to predict fitted values from the estimated model. Logistic regression is particularly valuable in that the predictions made from a fitted model are probabilities, constrained to be within the range of values 0–1. More accurately, a logistic regression model predicts the probability that the response has a value of 1 given a specific set of predictor values. Interpretation of logistic model coefficients usually involves their exponentiation, which allows them to be understood as odds ratios. This capability is unique to the class of logistic models, whether observation-based format or in grouped format. The fact that a logistic model can be used to assess the odds ratio of predictors, and also can be used to determine the probability of the response occurring based on specific predictor values, called covariate patterns, is the prime reason it has enjoyed such popularity in the statistical community for the past several decades.

1.3 THE BERNOULLI DISTRIBUTION

I have emphasized that binary response logistic regression is based on the Bernoulli probability distribution, which consists of a distribution of 1s and 0s. The probability function can be expressed as

$$f(y_i; p_i) = \prod_{i=1}^{n} p_i^{y_i} (1 - p_i)^{1-y_i} \tag{1.1}$$

where the PDF is the product, Π, of each observation in the data being modeled, symbolized by the subscript *i*. Usually the product term is dropped as being understood since all probability functions are products across the components of their respective distributions. We may then characterize the Bernoulli distribution as

$$f(y_i; p_i) = p_i^{y_i} (1 - p_i)^{1-y_i} \tag{1.2}$$

where *y* is the response variable being modeled and *p* is the probability that *y* has the value of 1. Again, 1 generally indicates a success, or that the event of interest has occurred. *y* only has values of 1 or 0, whereas *p* has values ranging from 0 to 1. p is many times symbolized as π or as μ. In fact, we shall be using the μ symbolization for the predicted mean or fit throughout most of the book.

A probability function generates or produces data on the basis of known parameters. That's the meaning of $f(y; p)$. What is needed in order to estimate the true parameters of the population data is to estimate the parameters on the basis of known data. After all we are modeling known data—and attempting to estimate parameter(s). We do this by inverting the order of y and p in the PDF. We attempt to calculate p on the basis of y. This relationship is called the *likelihood function.*

Statisticians may characterize the likelihood function as

$$f(p_i; y_i) = p_i^{y_i}(1 - p_i)^{1-y_i} \tag{1.3}$$

but usually parameterize the structure of the likelihood function by putting it into what is called exponential family form. Mathematically it is identical to Equation 1.3 above.

$$L(p_i; y_i) = \exp\left\{ y \ln\left(\frac{p_i}{1 - p_i}\right) + \ln(1 - p_i) \right\} \tag{1.4}$$

Note that "ln" is a symbol for the natural log of an expression. It is also symbolized as "log." Keep in mind that it differs from "log to the base 10," or "$\log_{10}$." The exponentiation of a logged value is the value itself; that is, $\exp(\ln(x)) = x$, or $e^{\ln(x)} = x$.

Statisticians usually take the log of both sides of the likelihood function, creating what is called the log-likelihood function. Doing this allows a summation across observations rather than multiplication. This makes it much easier for the algorithms used to estimate distribution parameters to converge; that is, to solve for the estimates. The Bernoulli log-likelihood function can be displayed as

$$\mathcal{L}(p_i; y_i) = \sum_{i=1}^{n} y \ln\left(\frac{p_i}{1 - p_i}\right) + \ln(1 - p_i) \tag{1.5}$$

The model parameters may be determined by simple calculus. Take the derivative of the log-likelihood function, set to 0, and solve. The problem is that the solution must be obtained iteratively, but for most all modeling situations, the solution takes only a few iterations to find the appropriate parameter estimates. The details of estimation can be found in Hilbe (2009). We need not worry about it in this discussion.

One of the nice features of presenting the log-likelihood function in exponential form is that we may abstract from it a link function as well as the mean and variance functions of the underlying Bernoulli distribution. The link function, which I'll discuss shortly, is whatever follows the y of the first term of the right-hand side of Equation 1.5. Here it is $\log(p/(1-p))$. The mean of the distribution can be obtained as the derivative of the negative of the second term with respect to the link. The second derivative yields the variance. For the Bernoulli distribution, these values are

$$\text{Mean} = \mu = p$$
$$\text{Variance} = V(\mu) = p(1-p) = \mu(1-\mu)$$

In the case of Bernoulli-based logistic regression, the mean is symbolized as μ (mu) and variance as $\mu(1-\mu)$. The above link and log-likelihood functions are many times expressed in terms of μ as well. It is important to note that strictly speaking the estimated p or μ should be symbolized as $\hat{p}$ and $\hat{\mu}$, respectively. p and μ are typically reserved for the true population values. However, for ease of interpretation, I will use the symbol μ in place of $\hat{\mu}$ throughout the book.

I should also mention that for grouped logistic regression, which we address in Chapter 5, μ and p are not the same, with μ defined as $n \cdot p$. But I'll delay making this distinction until we begin discussing grouped models.

Let us look at a logistic regression and how it differs from normal or ordinary linear regression. Recall that a regression attempts to understand a response variable on the basis of one of more predictors or explanatory variables. This is usually symbolized as

$$\hat{y}_i = \sum_{i=1}^{n} \beta_0 + x'_{i1}\beta_1 + x'_{i2}\beta_2 + \cdots + x'_{ij}\beta_j \tag{1.6}$$

where y-hat, or $\hat{y}$, is the sum of the terms in the regression. The sum of regression terms is also referred to as the linear predictor, or xb. Each $x'\beta$ is a term indicating the value of a predictor, x, and its coefficient, β. In linear regression, which is based in matrix form on the Gaussian or normal probability distribution, $\hat{y}$ is the predicted value of the regression model as well as the linear predictor. j indicates the number of predictors in a model. There is a linear relationship between the predicted or fitted values of the model and the terms on the right-hand side of Equation 1.6—the linear predictor. $\hat{y} = xb$. This is not the case for logistic regression.

The linear predictor of the logistic model is

$$x_i b = \sum_{i=1}^{n} \beta_0 + x'_{i1}\beta_1 + x'_{i2}\beta_2 + \cdots + x'_{ij}\beta_j \tag{1.7}$$

However, the fitted or predicted value of the logistic model is based on the link function, $\log(\mu/(1-\mu))$. In order to establish a linear relationship of the predicted value, μ, and the linear predictor, we have the following relationship:

$$\ln\left(\frac{\mu_i}{1-\mu_i}\right) = x_i b = \sum_{i=1}^{n} \beta_0 + x'_{i1}\beta_1 + x'_{i2}\beta_2 + \cdots + x'_{ij}\beta_j \tag{1.8}$$

where μ, like p, is the probability that the response value y is equal to 1. It can also be thought of as the probability of the presence or occurrence of some characteristic, while $1-p$ can be thought of as the absence of that characteristic. Notice that $\mu/(1-\mu)$, or $p/(1-p)$, is the formula for odds. The odds of something occurring is the probability of its success or presence divided by the probability of its failure or absence, $1-p$. If $\mu = 0.7$, $(1-\mu) = 0.3$. $\mu + (1-\mu)$ always equals 1. The log of the odds has been called by statisticians the *logit* function, from which the term logistic regression derives.

In order to determine μ on the basis of the linear predictor, xb, we solve the logit function for μ, without displaying subscripts, as

$$\mu = \frac{\exp(xb)}{1+\exp(xb)} = \frac{1}{1+\exp(-xb)} \tag{1.9}$$

These two forms of Equation 1.9 above are very important, which we will frequently use in our later discussion. Once a logistic model is solved, we may calculate the linear predictor and then apply the above formula to determine the predicted value for each observation in the model. μ is also the mean parameter. The parameter is shared by the individual observations of the model by μ.

1.4 METHODS OF ESTIMATION

Maximum likelihood estimation, MLE, is the standard method used by statisticians for estimating the parameter estimates of a logistic model. Other methods may be used as well, but some variety of MLE is used by nearly all statistical software for logistic regression. There is a subset of MLE though that can be used if the underlying model PDF is a member of the single parameter exponential family of distributions. The Bernoulli distribution is an exponential family member. As such, logistic regression can also be done from within the framework of generalized linear models or GLM. GLM allows for a much simplified manner of calculating parameter estimates, and is used in R with the *glm* function as the default method for logistic regression. It is a function in the R *stats* package, which is a base R package. Stata also has a *glm* command, providing the full range of GLM-based models, as well as full maximum likelihood estimate commands *logit* and *logistic*. The SAS *Genmod* procedure is a GLM-based procedure, and Proc *Logistic* is similar to Stata's *logit* and *logistic* commands. In Python, one may use the *statsmodels Logit* function for logistic regression.

Since R's default logistic regression is part of the *glm* function, we shall examine the basics of how it works. The *glm* function uses an iterative reweighted least squares (IRLS) algorithm to estimate the predictor coefficients of a logistic regression. The logic of a stand-alone R algorithm that can be used for logistic regression is given in Table 1.1. It is based on IRLS. I have annotated each line to assist in understanding how it works. You certainly do not have to understand the code to continue with the book. I have provided the code for those who are proficient in R programming. The code is adapted from Hilbe and Robinson (2013).

R users can paste the code from the table into the "New Script" editor. The code is an entire function titled *irls_logit*. The code is also available on the book's website, listed as *irls_logit.r*. Select the entire code, right click your mouse, and click on "Run line of selection." This places the code into active memory. To show what a logistic regression model looks like, we can load some data and execute the function. We shall use the *medpar* data set, which is 1991 Arizona inpatient Medicare (U.S. senior citizen national health plan) data. The data consist of cardiovascular disease patient information from a single diagnostic group. For privacy purposes, I did not disclose the diagnostic group to which the data are classified.

- los: length of stay (nights) in the hospital (continuous)
- hmo: 1 = patient a member of a Health Maintenance Organization; 0 = private pay
- white: 1 = patient identifies self as white; 0 = non-white
- died: 1 = patient dies within 48 hours of admission; 0 = did not die during this period
- age80: 1 = patient age 80 and over; 0 = less than 80
- type: type of admission—1 = elective; 2 = urgent; 3 = emergency

TABLE 1.1 R function for logistic regression

```
irls_logit <- function(formula, data, tol=.000001) {   # set option default values
  mf <- model.frame(formula, data)                     # define model frame as mf
  y <- model.response(mf, "numeric")                   # set model responses as y
  X <- model.matrix(formula, data = data)              # define X as matrix of data
  if (any(is.na(cbind(y, X)))) stop("Some data are missing.") # delete missing data
  mu <- (y + .5)/2                                     # initialize mu
  eta <- log(mu/(1-mu))                                # initialize linear predictor
  dev <- 2*sum(y*log(1/mu) + (1-y)*log(1/(1-mu)) )     # initialize deviance
  deltad <- 1                                          # initialize deltad
  i <- 1                                               # initialize i=1 iteraction log
 while (abs(deltad) > tol ) {                          # start IRLS loop
     w <- mu*(1-mu)                                    # weight - variance
     z <- eta + (y - mu)/w                             # working response
     mod <- lm(z ~ X-1, weights=w)                     # weighted linear regression
     eta <- mod$fit                                    # linear predictors from mod
     mu <- 1/(1 + exp(-eta))                           # fitted values; probabilities
     dev.old <- dev                                    # setup for convergence
     dev <- 2*sum(y*log(1/mu) + (1-y)*log(1/(1-mu)) )  # deviance - basis of convergence
     deltad <- dev - dev.old                           # calc difference new & old dev
     cat(i, coef(mod) , deltad, "\n")                  # iteration log
     i <- i+1                                          # update iteration number
  }                                                    # end IRLS loop
  beta <- mod$coef                                     # save coefficients
  pr <- sum(residuals(mod, type="pearson")^2)          # calc Pearson dispersion
  prdisp <- pr/mod$df.residual                         # calc Pearson dispersion
  return(list(coef = coef(mod),
         se = sqrt(diag(vcov(mod)))/sqrt(prdisp)))     # display of coef and SE
}                                                      # end irls_logit function
```

```
> library(LOGIT)
> data(medpar)
> head(medpar)
  los hmo white died age80 type provnum
1   4   0     1    0     0    1  030001
2   9   1     1    0     0    1  030001
3   3   1     1    1     1    1  030001
4   9   0     1    0     0    1  030001
5   1   0     1    1     1    1  030001
6   4   0     1    1     0    1  030001
```

We may run the model using the following code:

```
> mylogit <- irls_logit(died ~ hmo + hite, data=medpar)
> mylogit
$coef
X(Intercept)          Xhmo        Xwhite
 -0.92618620   -0.01224648    0.30338724

$se
X(Intercept)          Xhmo        Xwhite
   0.1973903     0.1489251     0.2051795
```

Just typing the model name we assigned, *mylogit*, displays the coefficients and standard errors of the model. We can make a table of estimates, standard errors, z-statistic, p-value, and confidence intervals by using the code:

```
> coef <- mylogit$coef
> se <- mylogit$se
> zscore <- coef / se
> pvalue <- 2*pnorm(abs(zscore),lower.tail=FALSE)
> loci <- coef - 1.96 * se
> upci <- coef + 1.96 * se
> coeftab <- data.frame(coef, se, zscore, pvalue, loci, upci)
> round(coeftab, 4)
                coef     se  zscore pvalue    loci    upci
X(Intercept) -0.9262 0.1974 -4.6922 0.0000 -1.3131 -0.5393
Xhmo         -0.0122 0.1489 -0.0822 0.9345 -0.3041  0.2796
Xwhite        0.3034 0.2052  1.4786 0.1392 -0.0988  0.7055
```

Running the same data using R's *glm* function produces the following output. I have deleted some ancillary output.

```
> glmlogit <- glm(died ~ hmo + white, family=binomial,
              data=medpar)
> summary(glmlogit)
                          . . .
Coefficients:
            Estimate Std. Error z value Pr(>|z|)
(Intercept) -0.92619    0.19739  -4.692  2.7e-06 ***
hmo         -0.01225    0.14893  -0.082    0.934
white        0.30339    0.20518   1.479    0.139
---
    Null deviance: 1922.9 on 1494 degrees of freedom
Residual deviance: 1920.6 on 1492 degrees of freedom
AIC: 1926.6
```

The confidence intervals must be calculated separately. To obtain model-based standard errors, we use the *confint.default* function. Using the *confint* function produces what are called profile confidence intervals. We shall discuss these later in Chapter 2, Section 2.3.

```
> confint.default(glmlogit)
                  2.5 %     97.5 %
(Intercept) -1.31306417 -0.5393082
hmo         -0.30413424  0.2796413
white       -0.09875728  0.7055318
```

Again, I have displayed a full logistic regression model output to show where we are headed in our discussion of logistic regression. The output is very similar to that of ordinary linear regression. Interpretation, however, is different. How coefficients, standard errors, and so forth are to be interpreted will concern us in the following chapters.

SAS CODE

```
/* Section 1.4 */
*Import medpar as a temporary dataset;
proc import datafile="c:\data\medpar.dta" out=medpar
dbms=dta replace;
```

```
run;

*Print the first six observations;
proc print data=medpar (obs=6);
run;

*Build the logistic model;
proc genmod data=medpar descending;
      model died=hmo white/ dist=binomial link=logit;
run;

*Another way to build the logistic model;
proc logistic data=medpar descending;
      model died=hmo white / clparm=both;
run;
```

STATA CODE

```
. use medpar
. glm died hmo white, fam(bin) nolog
```

Logistic Models *Single Predictor* 2

2.1 MODELS WITH A BINARY PREDICTOR

The simplest way to begin understanding logistic regression is to apply it to a single binary predictor. That is, the model we shall use will consist of a binary (0,1) response variable, y, and a binary (0,1) predictor, x. In addition, the data set we define will have 9 observations. Recall from linear regression that a response and predictor are paired when setting up a regression. Using R we assign various 1s and 0s to each y and x.

```
> y <- c(1,1,0,0,1,0,0,1,1)
> x <- c(0,1,1,1,0,0,1,0,1)
```

These values will be placed into a data set named *xdta*. Then we subject it to the *irls_logit* function displayed in the previous chapter.

```
> xdta <- data.frame(y,x)
> logit1 <- irls_logit(y ~ x, data=xdta)
```

The model name is *logit1*. Using the code to create the nice looking "standard" regression output that was shown before, we have

```
> coef <- logit1$coef
> se <- logit1$se
> zscore <- coef / se
> pvalue <- 2*pnorm(abs(zscore),lower.tail=FALSE)
> loci <- coef - 1.96 * se
```

```
> upci <- coef + 1.96 * se
> coeftab <- data.frame(coef, se, zscore, pvalue, loci, upci)
> round(coeftab, 4)
                 coef     se  zscore pvalue    loci   upci
X(Intercept) 1.0986 1.1547  0.9514 0.3414 -1.1646 3.3618
Xx          -1.5041 1.4720 -1.0218 0.3069 -4.3891 1.3810
```

The coefficient or slope of *x* is –1.5041 with a standard error of 1.472. The intercept value is 1.0986. The intercept is the value of the model when the value of *x* is zero.

Using R's *glm* function, the above data may be modeled using logistic regression as

```
> glm(y~ x, family = binomial, data = xdta)

Call: glm(formula = y ~ x, family = binomial, data = xdta)

Coefficients:
(Intercept)            x
     1.099       -1.504

Degrees of Freedom: 8 Total (i.e. Null); 7 Residual
Null Deviance:       12.37
Residual Deviance:   11.23       AIC: 15.23
```

More complete model results can be obtained by assigning the model a name, and then summarizing it with the *summary* function. We will name the model *logit2*.

```
> logit2 <- glm(y~ x, family = binomial, data = xdta)
> summary(logit2)

Call:
glm(formula = y ~ x, family = binomial, data = xdta)

Deviance Residuals:
      Min         1Q     Median        3Q       Max
  -1.6651    -1.0108     0.7585    0.7585    1.3537

Coefficients:
            Estimate Std. Error z value Pr(>|z|)
(Intercept)    1.099      1.155   0.951    0.341
x             -1.504      1.472  -1.022    0.307

(Dispersion parameter for binomial family taken to be 1)
```

```
    Null deviance: 12.365 on 8 degrees of freedom
Residual deviance: 11.229 on 7 degrees of freedom
AIC: 15.229
```

Model-based confidence intervals may be displayed by

```
> confint.default(logit2)
                2.5 % 97.5 %
(Intercept) -1.164557 3.361782
x           -4.389065 1.380910
```

A more efficient way of displaying a logistic regression using R is to encapsulate the summary function around the regression. It will be the way I typically display example results using R.

```
> summary(logit2 <- glm(y~ x, family=binomial, data=xdta))

                        . . .

Coefficients:
             Estimate Std. Error z value Pr(>|z|)
(Intercept)     1.099      1.155   0.951    0.341
x              -1.504      1.472  -1.022    0.307
                         . .
    Null deviance: 12.365 on 8 degrees of freedom
Residual deviance: 11.229 on 7 degrees of freedom
AIC: 15.229
```

There are a number of ancillary statistics which are associated with modeling data with logistic regression. I will show how to do this as we progress, and functions and scripts for all logistic statistics, fit tests, graphics, and tables are provided on the books web site, as well as in the LOGIT package that accompanies this book. The LOGIT package will also have the data, functions and scripts for the second edition of *Logistic Regression Models* (Hilbe, 2016).

For now we will focus on the meaning of the single binary predictor model. The coefficient of predictor x is −1.504077. A coefficient is a slope. It is the amount of the rate of change in y based on a one-unit change in x. When x is binary, it is the amount of change in y when x moves from 0 to 1 in value. But what is changed?

Recall that the linear predictor, xb, of a logistic model is defined as $\log(\mu/(1 - \mu))$. This expression is called the log-odds or logit. It is the logistic link function, and is the basis for interpreting logistic model coefficients. The interpretation of x is that when x changes from 0 to 1, the log-odds of

y changes by −1.504. This interpretation, although accurate, means little to most analysts.

What happens if we exponentiate $\log(\mu/(1 - \mu))$? The result is simply $\mu/(1 - \mu)$, which is interpreted as the odds of μ, with μ being the probability that $y = 1$, and $1 - \mu$ being the probability that $y = 0$ (the probability that y is not 1). By exponentiating the coefficient of x we may interpret the result as follows:

> The odds ratio of $x == 1$ is the ratio of the odds of $x = 1$ to the odds of $x = 0$.

The odds of $x = 1$ is exp(−1.504077) or 0.22222 times greater than the odds of $x = 0$. This is the same as saying that the odds of $x = 0$ is 1/exp(−1.504077) or 4.5 times greater than $x = 1$. The exponentiation of the intercept is not an odds ratio, but rather only an odds. Here it is exp(1.098612) or 3.0.

Another way of demonstrating this relationship is by constructing a table from the variables y and x.

```
> table(y,x)
   x
y   0 1
  0 1 3
  1 3 2
```

To add margin sums, use the code

```
> addmargins(table(y,x))
      x
y      0 1 Sum
  0    1 3   4
  1    3 2   5
  Sum  4 5   9
```

The odds of $x = 1$ is defined as “the value of $x = 1$ when $y = 1$ divided by the value of $x = 1$ when $y = 0$.” Here the odds of $x = 1$ is 2/3, or

Odds $x = 1$

```
> 2/3
[1] 0.6666667
```

The odds of $x = 0$ is,

Odds $x = 0$

```
. di 3/1
3
```

Creating a ratio of values we have

Odds Ratio $x = 1$ to $x = 0$

```
. di (2/3)/(3/1)
.22222222
```

That is…

To obtain the odds of $x = 1$: for $x = 1$, take the ratio of $y = 1$ to $y = 0$, or $2/3 = 0.666667$.
To obtain the odds of $x = 0$: for $x = 0$, take the ratio of $y = 1$ to $y = 0$, or $3/1 = 3$.

To obtain the odds ratio of $x = 1$ to $x = 0$, divide. Therefore, $0.666667/3 = 0.222222$
The intercept is the odds of $y = 1$ divided by $y = 0$ for $x = 0$, or 3.

The relationship of the logistic odds ratio and coefficient is:

$$\ln(\textit{Odds Ratio}) = \textit{coefficient}$$

$$\exp(\textit{coefficient}) = \textit{odds ratio}$$

Calculating the odds ratio and odds-intercept from the *logit2* model results,

Odds Ratio and Odds Intercept

```
> exp(logit2$coef)
(Intercept)           x
  3.0000000   0.2222222
```

Now we can reverse the relationships by taking the natural log of both.

Coefficient of x from Odds Ratio of x

```
> log(0.222222222)
[1] -1.504077
```

Intercept from Odds of Intercept

```
> log(3)
[1] 1.098612
```

2.2 PREDICTIONS, PROBABILITIES, AND ODDS RATIOS

I mentioned before that unlike linear regression, the model linear predictors and fitted values differ for logistic regression. If μ is understood as the predicted mean, or fitted value:

Linear regression	$\mu = x'\beta$
Logistic regression	$\mu = \exp(x'\beta)/(1 + \exp(x'\beta))$
or	$\mu = 1/(1 + \exp(-x'\beta))$

For the logistic model, μ is defined as the probability that $y = 1$, where y is the symbol for the model response term.

```
> logit2 <- glm( y ~ x, family=binomial, data=xdta)

> coef(logit2)
(Intercept)           x
   1.098612   -1.504077

LINEAR PREDICTOR WHEN X=1
> 1.098612 -1.504077*1
[1] -0.405465

LINEAR PREDICTOR WHEN X=0
> 1.098612 -1.504077*0
[1] 1.098612
```

We use R's post-*glm* function for calculating the linear predictor. The code below generates linear predictor values for all observations in the model. Remember that R has several ways that certain important statistics can be obtained.

```
> xb <- logit2$linear.predictors
```

The inverse logistic link function is used to calculate μ.

```
> mu <- 1/(1+exp(-xb))
```

From the predicted probability that $y = 1$, or μ, the odds for each level of x may be calculated.

```
> o <- mu/(1-mu)
```

Let us now check the relationship of x to o, noting the values of o for the two values of x.

```
> check_o <-data.frame(x,o)
> round(check_o, 3)
  x     o
1 0 3.000
2 1 0.667
3 1 0.667
4 1 0.667
5 0 3.000
6 0 3.000
7 1 0.667
8 0 3.000
9 1 0.667
```

Recall that the odds ratio of x is the ratio of $x = 1/x = 0$. The odds of the intercept is the value of o when $x = 0$. In order to obtain the odds ratio of x when $x = 1$, we divide 0.667/3. So that we do not have rounding problems with the calculations, $o = 0.667$ will be indicated as $o < 1$. We will create a variable called *or* that retains the odds-intercept value ($x = 0$) or 3.0 and selectively changes each value of $o < 1$ to 0.667/3. The corresponding model coefficient may be determined by logging each value of *or*.

```
> or <- o
> or[or< 1] <- (.6666667/3)
> coeff <- log(or)
```

Finally we shall create a table of statistics, including all of the relevant values we have just calculated.

```
> data1 <-data.frame(y,x,xb,mu,o,or,coeff)
> round(data1,4)
  y x      xb   mu      o     or   coeff
1 1 0  1.0986 0.75 3.0000 3.0000  1.0986
2 1 1 -0.4055 0.40 0.6667 0.2222 -1.5041
3 0 1 -0.4055 0.40 0.6667 0.2222 -1.5041
4 0 1 -0.4055 0.40 0.6667 0.2222 -1.5041
5 1 0  1.0986 0.75 3.0000 3.0000  1.0986
6 0 0  1.0986 0.75 3.0000 3.0000  1.0986
7 0 1 -0.4055 0.40 0.6667 0.2222 -1.5041
8 1 0  1.0986 0.75 3.0000 3.0000  1.0986
9 1 1 -0.4055 0.40 0.6667 0.2222 -1.5041
```

What we find is that from the model linear predictor and probabilities we calculated the model odds ratios and coefficients. Adding additional predictors and formatting predictors as categorical and continuous allow us to do the same thing as we did for a single binary predictor—it is just a bit more complex. See Hilbe (2016) or the PDF document, "Calculating Odds Ratios from Probabilities" on the author's web site for the book. My goal here is to demonstrate how odds, odds ratios, coefficients and probabilities relate with one another in a logistic model. You can also understand why the model coefficients are referred to as parameter estimates. Each coefficient contributes to the mean parameter being estimated by the logistic model. Likewise, we may also see how the fitted values, or probabilities, all relate as components of the mean parameter estimated by the model.

2.3 BASIC MODEL STATISTICS

The output provided by most statistical software for logistic regression involves a display of basic model statistics as well as several important statistics that are important for assessing model fit. The basic model statistics nearly always include the model intercept, one or more coefficients and associated standard errors, z statistics, p-values, and confidence intervals. Exponentiated logistic coefficients are referred to as odds ratios. I refer to the exponentiated intercept as the odds-intercept.

We have already displayed all of these statistics and calculated each by hand using R software. However, except for coefficients and odds ratios little has thus far been said about them.

R's *glm* function utilizes a *summary* function to display logistic model coefficients/odds ratios, standard errors, z statistics, and p-values. A separate function is required to obtain confidence intervals. Model-based confidence intervals are calculated using *confint.default*(), but the preferred way of producing confidence intervals with *glm* is by use of the *confint*() function. As we shall discuss later, *confint* calculates *profile confidence intervals.* These are much more complicated to calculate, but are definitely to be preferred over simple model-based intervals.

2.3.1 Standard Errors

Standard errors provide the analyst with information concerning the variability of the coefficient. If a coefficient is an estimate of the true coefficient or slope that exists within the underlying probability distribution describing the data being analyzed, then the standard error tells us about the accuracy of the "point" estimate of the coefficient. Essentially it allows us to determine if the coefficient is significantly different from 0. A coefficient of 0 indicates no effect, and contributes nothing to understanding the response variable of interest.

On the basis of the maximum likelihood estimates, the standard errors derive from the negative inverse Hessian matrix, or the second derivatives of the log-likelihood function. Specifically, the standard errors are the square roots of the diagonal elements of the model negative inverse Hessian matrix. This matrix is also commonly referred to by analysts as the variance–covariance matrix. It can be obtained in R by using the *vcov* function

```
> summary(logit2 <- glm(y ~ x, family=bin, data=xdta)))

                              . . .

Coefficients:
              Estimate Std. Error z value Pr(>|z|)
(Intercept)      1.099      1.155   0.951    0.341
x               -1.504      1.472  -1.022    0.307

                              . . .

> vcov(logit2)
            (Intercept)         x
(Intercept)    1.333331 -1.333331
x             -1.333331  2.166664
```

The diagonal elements are 1.3333 for the intercept and 2.16666 for predictor *x*. These are the variances of the intercept and of *x*.

```
> diag(vcov(logit2))
(Intercept)         x
   1.333331  2.166664
```

Taking the square root of the variances gives us the model standard errors.

```
> sqrt(diag(vcov(logit2)))
(Intercept)         x
   1.154700  1.471959
```

These values are identical to the standard errors shown in the *logit2* results table. Note that when using R's *glm* function, the only feasible way to calculate model standard errors is by use of the `sqrt(diag(vcov(modelname)))` method. The `modelname$se` call made following the *irls_logit* function from Table 1.1 cannot be used with *glm*.

Analysts many times make adjustments to model standard errors when they suspect excess correlation in the data. Correlation can be derived from a variety of sources. One of the earliest adjustments made to standard errors was called *scaling*. R's *glm* function provides built in scaling of binomial and Poisson regression standard errors through the use of the *quasibinomial* and *quasipoisson* options. Scaled standard errors are produced as the product of the model standard errors and square root of the Pearson dispersion statistic. Coefficients are left unchanged. Scaling is discussed in detail in Chapter 3, Section 3.4.1.

```
> summary(logitsc <- glm( y ~ x, family=quasibinomial, data=xdta))

Coefficients:
            Estimate Std. Error t value Pr(>|t|)
(Intercept)      1.099      1.309     0.839     0.429
x               -1.504      1.669    -0.901     0.397

(Dispersion parameter for quasibinomial family taken to be
1.285715)
```

I will explain more about the Pearson statistic, the Pearson dispersion, scaling, and other ways of adjusting standard errors when we discuss model fit. However, it is easy to observe that the scaled standard error for x in model *logit2* above is calculated by

```
> 1.471959 * sqrt(1.285715)
[1] 1.669045
```

based on the formula I described. The dispersion statistic is displayed in the final line of the quasibinomial model output above. Regardless, many analysts advise that standard errors be adjusted by default. If data are not excessively correlated, scaled standard errors, for example, reduce to model standard errors.

The standard errors of odds ratios cannot be abstracted from a variance–covariance matrix. One calculates odds ratio standard errors using what statisticians call the *delta* method. See Hilbe (2009, 2016) for details. When the *delta* method is used for odds ratios, as well as for risk or rate ratios, the calculation is simple.

$$SE_{OR} = exp(\beta)*SE_{coef}$$

Standard errors of odds ratios are calculated by multiplying the odds ratio by the coefficient standard error. Starting from the *logit2* model, odds ratios and their corresponding standard errors maybe calculated by,

```
> logit2 <- glm( y ~ x, family=binomial, data=xdta)
> coef <- logit2$coefficients              # coefficients
> or <- exp(logit2$coefficients)           # odds ratios
```

```
> se <- sqrt(diag(vcov(logit2)))  # coefficient SE
> delta <- or*se                  # delta method,SE of OR
> ortab <- data.frame(or, delta)
> round(ortab, 4)
                or   delta
(Intercept) 3.0000 3.4641
x           0.2222 0.3271
```

2.3.2 *z* Statistics

The z statistic is the ratio of a coefficient to its standard error.

```
> zscore <- coef/se
```

The reason this statistic is called z is due to its assumption as being normally distributed. For linear regression models, we use the t statistic instead. The z statistic for odds ratio models is identical to that of standard coefficient models. Large values of z typically indicate a predictor that significantly contributes to the model; that is, to the understanding of the response.

2.3.3 *p*-Values

The p-value of a logistic model is usually misinterpreted. It is also typically given more credence than it should have. First, though, let us look at how it is calculated.

```
> pvalue <- 2*pnorm(abs(zscore),lower.tail=FALSE)
```

The p-value is a two-tail test of the z statistic. It tests the null hypothesis that the associated coefficient value is 0. More exactly, p is the probability of obtaining a coefficient value at least as great as the observed coefficient given that the assumption that $\beta = 0$. The smaller the p-value, the greater the probability that $\beta \neq 0$. The standard "level of significance" for most studies is $p = 0.05$. Values of less than 0.05 indicate that the null hypothesis of no relationship between the predictor and response is false. That is, p-values less than 0.05 indicate that the predictor significantly contributes to the model. Values greater than 0.05 indicate that the null hypothesis has not been rejected and that the predictor does not contribute to the model.

A cutoff of 0.05 means that one out of every 20 times the coefficient on average will not reject the null hypothesis; that is, that the coefficient is in fact not significant when we thought it was. For many scientific disciplines, this is not a strict enough criterion. In astrostatistics, for example, preferred criteria of statistical significance range from 0.01 to 0.001. There are a number of issues related to power and false positives when discussing an appropriate criterion for a p-value. Determine what criterion makes sense for the type of study in which you are engaged rather than simply apply a 0.05 criterion without question.

Regression software assumes that the p-value is based on a two-tailed test. In some studies a one-tailed test is more appropriate. You predict that the direction of the coefficient in question goes only in one direction. When this is the case be sure to divide the displayed p-value by 2 prior to assessing its significance.

2.3.4 Confidence Intervals

Model based 95% confidence intervals are calculated as follows:

```
> loci <- coef - qnorm(.975) * se
> upci <- coef + qnorm(.975) * se
```

where *qnorm* is the outside 2.5% of the observations from each side of the normal distribution.

```
> qnorm(.975)
[1] 1.959964
```

Together the distribution excludes 5% of the distribution, or 0.05. Many times analysts will use 1.96 instead of the *qnorm* function when calculating confidence intervals. The confidence intervals for odds ratios are exponentiations of the coefficient-based confidence intervals. Combining everything together, we can use the code below to produce a table displaying the odds ratio of x and the odds-intercept together with their related standard errors, z statistics, p-values, and confidence intervals. I should mention that the model-based confidence intervals we have been discussing are also referred to as Wald confidence intervals.

Calculation of Odds Ratio and Associated Model Statistics

```
> coef <- logit2$coef
> se <- sqrt(diag(vcov(logit2)))
> zscore <- coef / se
> or <- exp(coef)
> delta <- or * se
> pvalue <- 2*pnorm(abs(zscore),lower.tail=FALSE)
```

```
> loci <- coef - qnorm(.975) * se
> upci <- coef+qnorm(.975) * se
> ortab <- data.frame(or, delta, zscore, pvalue, exp(loci),
                     exp(upci))
> round(ortab, 4)
                  or delta zscore pvalue exp.loci. exp.upci.
(Intercept) 3.0000 3.4641 0.9514 0.3414 0.3121 28.8405
x           0.2222 0.3271 -1.0218 0.3069 0.0124 3.9785
```

Unfortunately, R users must program a table of odds ratio statistics as I have above. The *summary* function following *glm* displays a coefficient table of logistic model base statistics, but there is no function that automatically displays a table of odds ratio statistics. We can create an R function to do just. We shall call it *toOR.R* (see Table 2.1), where the OR component of *toOR* must be in capitals. R is case sensitive. I will place the *toOR* function into the LOGIT package so that it can be used automatically anytime the package is installed and loaded into memory. It can be used following the use of *glm* with the binomial family and default logit function. I will also place the function on the book's web site.

After estimation of a logistic regression using *glm*—for example, the *logit2* model—type

```
> toOR(logit2)

                or  delta  zscore pvalue exp.loci.  exp.upci.
(Intercept) 3.0000 3.4641  0.9514 0.3414    0.3121    28.8405
x           0.2222 0.3271 -1.0218 0.3069    0.0124     3.9785
```

TABLE 2.1 *toOR* function

```
toOR <- function(object, ...) {
     coef <- object$coef
       se <- sqrt(diag(vcov(object)))
       zscore <- coef / se
     or <- exp(coef)
       delta <- or * se
     pvalue <- 2*pnorm(abs(zscore),lower.tail=FALSE)
     loci <- coef - qnorm(.975) * se
     upci <- coef+qnorm(.975) * se
     ortab <- data.frame(or, delta, zscore, pvalue,
      exp(loci), exp(upci))
     round(ortab, 4)
}
```

We may use the function on *medpar* data

```
> data(medpar) # assumes library(COUNT)or library(LOGIT) loaded
> smlogit <- glm(died ~ white + los + factor(type),
              family = binomial, data = medpar)
> summary(smlogit)
```

. . .

```
Coefficients:
                 Estimate Std. Error z value Pr(>|z|)
(Intercept)     -0.716364   0.218040  -3.285  0.00102 **
white            0.305238   0.208926   1.461  0.14402
los             -0.037226   0.007797  -4.775  1.80e-06 ***
factor(type)2    0.416257   0.144034   2.890  0.00385 **
factor(type)3    0.929994   0.228411   4.072  4.67e-05 ***

> toOR(smlogit)
                    or  delta  zscore pvalue exp.loci. exp.upci.
(Intercept)     0.4885 0.1065 -3.2855 0.0010    0.3186    0.7490
white           1.3569 0.2835  1.4610 0.1440    0.9010    2.0436
los             0.9635 0.0075 -4.7747 0.0000    0.9488    0.9783
factor(type)2 1.5163 0.2184  2.8900 0.0039    1.1433    2.0109
factor(type)3 2.5345 0.5789  4.0716 0.0000    1.6198    3.9657
```

Confidence intervals are very important when interpreting a logistic model, as well as any regression model. By looking at the low and high range of a predictor's confidence interval an analyst can determine if the predictor contributes to the model.

Remember that a regression p-value is an assessment of whether we may "significantly" reject the null hypothesis that the coefficient (β) is equal to 0. If the confidence interval of a predictor includes 0, then we cannot be significantly sure that the coefficient is not really 0 in value. For odds ratios, since the confidence intervals are exponentiations of the coefficient confidence intervals, having the range of the confidence interval include 1 is evidence that the null hypothesis has not been rejected. The confidence intervals for *logit2* odds ratio model above both include 1—0.0124123 to 3.9788526 and 0.3120602 to 28.84059. Note that the p-values for both x and the intercept are approximately 0.3. 0.3 far exceeds the 0.05 criterion of significance.

How is the confidence interval to be interpreted? If zero is not within the lower and upper limits of the confidence interval of a coefficient, we cannot conclude that we are 95% sure that the coefficient is "significant"; that is, that the associated p-value is truly under 0.05. Many analysts interpret confidence intervals in such a manner, but they should not.

The traditional logistic regression model we are discussing here is based on a frequency interpretation of statistics. As such the confidence intervals must be interpreted in the same manner. If the coefficient of a logistic model predictor has a p-value under 0.05, the associated confidence interval will not include zero. The interpretation is

> *Wald Confidence Intervals*
> *If we repeat the modeling analysis a very large number of times, the true coefficient would be within the range of the lower and upper levels of the confidence interval 95 times out of 100.*

I earlier mentioned that the use of *confint*() following R's *glm* displays profile confidence intervals. *confint.default*() produces standard confidence intervals, based on the normal distribution. Profile confidence intervals are based on the *Chi*2 distribution. Profile confidence intervals are particularly important to use when there are relatively few observations in the model, as well as when the data are unbalanced. For example, if a logistic model has 30 observations, but the response variable consists of 26 ones and only 4 zeros, the data are unbalanced. Ideally a logistic response variable should have relatively equal numbers of 1s to 0s. Likewise, if a binary predictor has nearly all 1s or 0s, the model is unbalanced, and adjustments may need to be made to the model.

In any case, profile confidence intervals are derived as the inverse of the likelihood ratio test defined as

$$\text{Likelihood ratio test} = -2\{\mathcal{L}_{\text{reduced}} - \mathcal{L}_{\text{full}}\}$$

This is a test we will use later when assessing the significance of adding, or dropping, a predictor or group of predictors from a model. The log-likelihood of a model with all of the predictors is subtracted from the log-likelihood of a model with fewer predictors. The result is multiplied by " – 2." The significance of the test is based on the *Chi*2 distribution, whose arguments are the likelihood ratio test statistic and degrees of freedom. The degrees of freedom consists of how many predictors there are between the full and reduced models. If a single predictor is being evaluated, there is one degree of freedom. The likelihood ratio test is preferred to the standard Wald assessment based on regression coefficient or odds ratio p-values. We shall discuss the test further in Chapter 4, Section 4.2.

For now you need only know that profile confidence intervals are the inversion of the likelihood ratio test. A range of values for a coefficient, β, are produced for which the null hypothesis of $\beta = 0$ would not be rejected. The statistic is not simple to produce by hand, but easy to display using the *confint* function. It should be noted that when the predictors are significant and the logistic model is well fit, Wald or model-based confidence intervals differ little from profile confidence intervals. In the case of the *logit2* model where neither *x* nor the intercept are significant and there are only nine observations in the model, we expect for there to be a somewhat substantial difference in confidence interval values.

Wald or Model-Based Confidence Intervals

```
> confint.default(logit2)
                 2.5 %   97.5 %
(Intercept) -1.164557 3.361782
x           -4.389065 1.380910
```

Profile Confidence Intervals

```
> confint(logit2)
Waiting for profiling to be done...
                  2.5 %   97.5 %
(Intercept) -0.9568748 4.105099
x           -4.9264210 1.219928
```

Stata's *pllf* command produces profile confidence intervals, but only for continuous predictors.

Scaled, sandwich or robust, and bootstrapped-based confidence intervals will be discussed in Chapter 4, and compared with profile confidence intervals. We shall discuss which should be used given a particular type of data.

2.4 MODELS WITH A CATEGORICAL PREDICTOR

For our discussion of a logistic model with a single categorical predictor I shall return to the *medpar* data described in Chapter 1. I provided an introductory logistic model of *died* on *white* and *hmo*, which are all binary variables. *Type*, on the other hand, is a categorical variable with three levels. As indicated earlier, *type* = 1 signifies a patient who electively chose to be admitted to a hospital, *type* = 2 is used for patients who were admitted to the hospital as "urgent," and *type* = 3 is reserved for those patients who were admitted as emergency. *provnum* is a string variable designating the hospital provider number of the patients whose data are given in the respective lines or observations. I will use only *died* (1 = died within 48 h of admission) and *type* in this section.

```
> library(COUNT)
> data(medpar)
> head(medpar)
  los hmo white died age80 type provnum
1   4   0     1    0     0    1  030001
2   9   1     1    0     0    1  030001
3   3   1     1    1     1    1  030001
4   9   0     1    0     0    1  030001
5   1   0     1    1     1    1  030001
6   4   0     1    1     0    1  030001
```

We can check how many are in each level of *type*

```
> table(medpar$type)

   1    2    3
1134  265   96
```

and we can find out the percentage in each level

```
> prop.table( table(medpar$type))

         1          2          3
0.75852843 0.17725753 0.06421405
```

A no-frills frequency table may be produced by a little programming.

```
> Cnt <- table(medpar$type)
> Freq <- prop.table( table(medpar$type))
> typetab <- data.frame(Cnt, Freq)
> my1 <- typetab[ ,1:2]
> Pct <- typetab[ ,4]
> data.frame(my1, Pct)

  Var1 Freq        Pct
1    1 1134 0.75852843
2    2  265 0.17725753
3    3   96 0.06421405
```

Statisticians have handled categorical predictors in a variety of ways. When used with regression, and in particular with logistic regression, categorical predictors are nearly always factored into separate indicator or dummy variables. Each indicator variable has a value of 1 or 0 except for the reference level, which is excluded from the regression.

Think of each level except the reference as the $x = 1$ level with the reference variable as $x = 0$. If level 1 of a categorical predictor is taken as the reference level, then level 2 is interpreted with reference to level 1. Level 3 is also interpreted with reference to level 1. Level 1 is the default reference level for both R's *glm* function and Stata's regression commands. SAS uses the highest level as the default reference. Here it would be level 3.

It is advised to use either the lowest or highest level as the reference, in particular whichever of the two has the most observations. But of even more importance, the reference level should be chosen which makes most sense for the data being modeled.

You may let the software define your levels, or you may create them yourself. If there is the likelihood that levels may have to be combined, then it may be wise to create separate indicator variables for the levels. First though, let us let the software create internal indicator variables, which are dropped at the conclusion of the display to screen.

```
> summary(logit3 <- glm( died ~ factor(type), family=binomial,
          data=medpar))

                              . . .

Coefficients:
                Estimate Std. Error z value Pr(>|z|)
(Intercept)     -0.74924    0.06361 -11.779  < 2e-16 ***
factor(type)2    0.31222    0.14097   2.215 0.02677 *
factor(type)3    0.62407    0.21419   2.914 0.00357 **
--
    Null deviance: 1922.9 on 1494 degrees of freedom
Residual deviance: 1911.1 on 1492 degrees of freedom
AIC: 1917.1
```

Note how the *factor* function excluded *factor*(*type*1) (elective) from the output. It is the reference level though and is used to interpret both *type*2 (urgent) and *type*3 (emergency). I shall exponentiate the coefficients of *type*2 and *type*3 in order to better interpret the model. Both will be interpreted as odds ratios, with the denominator of the ratio being the reference level.

```
> exp(coef(logit3))
  (Intercept) factor(type)2 factor(type)3
    0.4727273     1.3664596     1.8665158
```

The interpretation is

- Urgent admission patients have a near 37% greater odds of dying in the hospital within 48 h of admission than do elective admissions.
- Emergency admission patients have a near 87% greater odds of dying in the hospital within 48 h of admission than do elective admissions.

Analysts many times find that they must change the reference levels of a categorical predictor. This may be done with the following code. We will change from the default reference level 1 to a reference level 3 using the *relevel* function.

```
> medpar$type <- factor(medpar$type)
> medpar$type <- relevel(medpar$type, ref=3)
> logit4 <- glm( died~factor(type), family=binomial,
                 data=medpar)
> exp(coef(logit4))
  (Intercept) factor(type)1 factor(type)2
    0.8823529     0.5357576     0.7320911
```

Interpretation changes to read

- Elective patients have about half the odds of dying within 48 h of admission to the hospital than do emergency patients.
- Urgent patients have about a quarter of the odds of dying within 48 h of admission to the hospital than do emergency patients.

I mentioned that indicator or dummy variables can be created by hand, and levels merged if necessary. This occurs when, for example, the level 2 coefficient (or odds ratio) is not significant compared to reference level 1. We see this with the model where *type* = 3 is the reference level. From looking at the models, it appears that levels 2 and 3 may not be statistically different from one another, and may be merged. I caution you from concluding this though since we may want to adjust the standard errors, resulting in changed *p*-values, for extra correlation in the data, or for some other reason we shall discuss in Chapter 4. However, on the surface it appears that patients who were admitted as urgent are not significantly different from emergency patients with respect to death within 48 h of admission.

I mentioned before that combining levels is required if two levels do not significantly differ from one another. In fact, when the emergency level of *type* is the reference, level 2 (urgent) does not appear to be significant, indicating that *type* levels 2 and 3 might be combined. With R this can be done as

```
> table(medpar$type)

   1    2    3
1134  265   96

> medpar$type[medpar$type==3] <- 2 # reclassify level 3 as level 2
> table(medpar$type)
```

```
   1    2
1134  361

> summary(logit6 <- glm(died~ factor(type), family=binomial,
        data=medpar))

                          . . .

Coefficients:
                 Estimate Std. Error z value Pr(>|z|)
(Intercept)      -0.74924    0.06361 -11.779  <2e-16 ***
factor(type)2     0.39660    0.12440   3.188  0.00143 **
```

2.5 MODELS WITH A CONTINUOUS PREDICTOR

2.5.1 Varieties of Continuous Predictors

A continuous predictor can take negative as well as zero and positive numeric values. Continuous predictors cause more problems for analysts than do discrete predictors; that is, binary and categorical predictors. The distribution or shape that a continuous variable takes may not be appropriate for including in a logistic model unless it is transformed in some manner. The key concept to remember is that a continuous predictor must be what is referred to by statisticians as "linear in the logit." This means that the continuous predictor needs to have a linear relationship with the logistic link function, $\log(\mu/(1 - \mu))$. A variable that is highly curved will not generally be linear in the logit.

Another feature of a continuous predictor to check is its range. If the predictor is, for example, years of age from 21 to 65, it is better to center or even standardize it. We will discuss these operations and their rationale later in this chapter. I mentioned these two problems areas related to continuous because they seem to cause problems for many analysts.

One of the foremost problems in dealing with continuous predictors in regression models has to do with the fact that a single coefficient represents the entire range of values. Recall that a regression coefficient is nothing more than a slope; that is, the rate of change in the response for a one-unit change in the predictor. This rate of change is assumed to be the same at any point in the variable. We can assume this since the logistic link function linearizes the relationship between the linear predictor and fitted value.

What happens when a predictor is curved like a parabola? Analysts typically transform the variable by squaring it, and entering both the original variable and squared variable in the model. Other transforms are commonly applied to continuous predictors including the square root, inverse, inverse square, and log. Probably the most common transform is the log transform.

There are downsides when transforming a continuous predictor. The major problem is interpretability. We must incorporate the transform made to a variable into its interpretation. For a log transform, we need to affirm that for a one-unit change in the predictor the response changes by the log of the response. Some analysts apply complex transforms to straighten out or linearize the relationship between the logit and predictor. But when it comes time to interpreting the meaning of the coefficient they are at a loss.

Linearizing a predictor in the context of logistic regression is more difficult than it is for linear regression where the linear predictor and fitted value are identical. In addition, other predictors in the model may affect the relationship of the fit and predictor. Partial residual plots and the use of generalized additive models (GAMs) are typically used to assess the best way to transform a continuous predictor. We shall discuss these tests in more detail in Chapter 3.

The interpretation of a continuous predictor is based on the same logit as for binary and categorical predictors. An odds ratio is the ratio of the odds when $x = 1$ to the odds when $x = 0$. For a binary predictor this is simple. For a multilevel categorical predictor, the reference level is the $x = 0$ level. For a continuous predictor, the lower value of two contiguous values in the predictor is the reference; the higher is the $x = 1$ level. For an age predictor, calculate the odds of age = 21 compared to the odds of age 20. The ratio odds(21)/odds(20) is the odds ratio, which is the same value for all pairs of values in the predictor. If the odds ratio for age is 1.01 and response is died, we can assert that the odds of death is 1% greater for each 1 year greater age of a patient.

2.5.2 A Simple GAM

I mentioned before, when including a continuous predictor into a logistic model, it is assumed that the slope or rate of change in the response for a one-unit change in the predictor is the same throughout the entire range of predictor values. It may be preferred to factor a continuous predictor at the points where its slope changes in any substantial manner. There is loss of information when this is done, but there is perhaps a gain in accuracy when interpreting the coefficient. GAMs is a widely used method to check the underling shape of a continuous predictor, adjusted by other predictors (none here), within the framework of a particular GLM family model: for example, logistic regression. For an example let us evaluate the variable *los* in the *medpar* data. LOS is an acronym for Length of Stay, referring to nights in the hospital. *los* ranges from 1 to 116. A cubic spline is used to smooth the shape of the distribution of *los*. This is accomplished by using the S operator.

```
> summary(medpar$los)
   Min. 1st Qu.  Median    Mean 3rd Qu.    Max.
  1.000   4.000   8.000   9.854  13.000 116.000

> library(mgcv)
> diedgam <- gam(died ~ s(los), family=binomial, data=medpar)
> summary(diedgam)

                                   . . .

Parametric coefficients:
            Estimate Std. Error z value Pr(>|z|)
(Intercept) -0.69195    0.05733  -12.07   <2e-16 ***

Approximate significance of smooth terms:
         edf Ref.df Chi.sq p-value
s(los) 7.424  8.292  116.8  <2e-16 ***

R-sq.(adj) = 0.0873 Deviance explained=6.75%
UBRE = 0.21064 Scale est. = 1 n = 1495
> plot(diedgam)
```

Note that no other predictors are in this model. Adding others may well alter the shape of the splines. The *edf* statistic indicates the "effective degrees of freedom." It is a value that determines the shape of the curves. An *edf* of 1 indicates a straight line; 8 and higher is a highly curved shape. The graph has an *edf* of 7.424, which is rather high. See Zuur (2012) for a complete analysis of GAM using R.

If this was all the data I had to work with, based on the change of slope points in Figure 2.1, I would be tempted to factor *los* into four intervals with three slopes at 10, 52, and 90. Each of the four levels would be part of a categorical predictor with the lowest level as the reference. If the slopes differ considerably across levels, we should use it for modeling the effect of *los* rather than model the continuous predictor.

2.5.3 Centering

A continuous predictor whose lowest value is not close to 0 should likely be centered. For example, we use the *badhealth* data from the COUNT package.

```
> data(badhealth)
> head(badhealth)
```

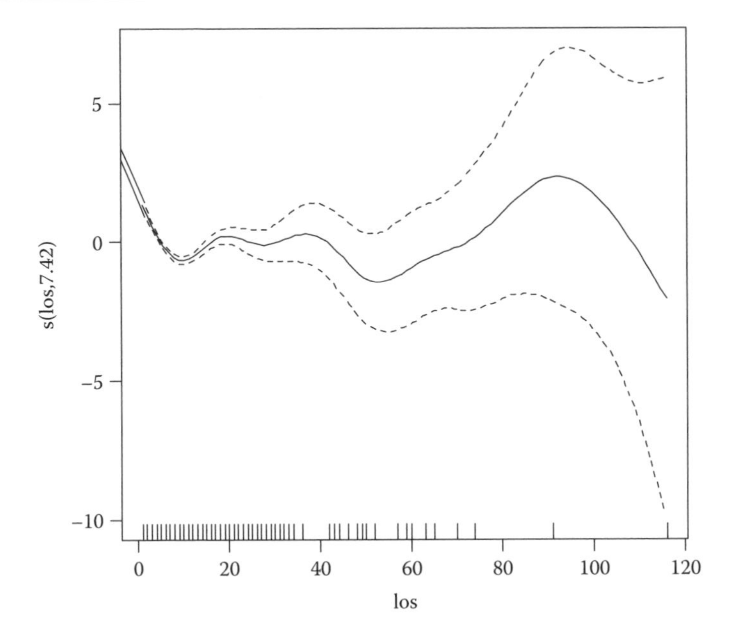

FIGURE 2.1 GAM model of *los*.

```
  numvisit badh age
1       30    0  58
2       20    0  54
3       16    0  44
4       20    0  57
5       15    0  33
6       15    0  28
```

badh is a binary variable, and indicates that a patient has "bad health," whatever that may mean. *numvisit*, or number of visits to the doctor during the year 1984, and *age*, are continuous variables. Number of *visits* ranges from 0 to 40, and the *age* range of patients is from 20 to 60.

```
> table(badhealth$badh)

   0    1
1015  112

> summary(badhealth$age)
   Min. 1st Qu.  Median    Mean 3rd Qu.    Max.
  20.00   28.00   35.00   37.23   46.00   60.00
```

```
> summary(badhealth$numvisit)
   Min. 1st Qu.  Median    Mean 3rd Qu.    Max.
  0.000   0.000   1.000   2.353   3.000  40.000
```

Centering allows a better interpretation for a predictor like *age*. Centering a continuous predictor with its low value not close to 0 is recommended when the variable is used in an interaction, as well as when it is used in a Bayesian model.

Centering is accomplished by subtracting the mean of the variable from each value of the variable. That is:

Centering: $x_i - mean(x_i)$

```
> cage <- badhealth$age - mean(badhealth$age)
> summary(cage)
   Min. 1st Qu.  Median    Mean 3rd Qu.    Max.
-17.230  -9.229  -2.229   0.000   8.771  22.770
```

The same result can be determined by use of the *scale* function.

```
> cenage <- scale(badhealth$age, center=TRUE, scale=FALSE)
```

Comparing the coefficients for models with *age* and centered age (*cage*):

```
> bad1 <- glm(badh ~ age, family=binomial, data=badhealth)
> bad2 <- glm(badh ~ cage, family=binomial, data=badhealth)
> badtab <- data.frame(bad1$coefficients, bad2$coefficients)
> badtab
            bad1.coefficients bad2.coefficients
(Intercept)       -4.58866278       -2.37171785
age                0.05954899        0.05954899
```

2.5.4 Standardization

Standardization of continuous predictors is important when other continuous predictors in your model are recorded on entirely different scales. The way this is done is by dividing the centered variable by the variable standard deviation. Use of R's scale function makes this easy:

```
> sage <- scale(badhealth$age)
> bad3 <- glm(badh ~ sage, family=binomial, data=badhealth)
```

The standard, centered, and standardized coefficient values for the *badhealth* data may be summarized in the following table. The intercept changes when a predictor such as *age* is centered. When a predictor is standardized both the intercept and predictor coefficients are changed with respect to a standard model. Note that the intercept remains the same when a predictor is either centered or standardized.

```
> badtab2 <- data.frame(bad1$coefficients,
bad2$coefficients, bad3$coefficients)
> badtab2
            bad1.coefficients bad2.coefficients bad3.coefficients
(Intercept)       -4.58866278       -2.37171785        -2.3717178
age                0.05954899        0.05954899         0.6448512
```

Again, standardization is warranted when two or more continuous predictors in a model are measured on different scales, making it difficult to compare them. Interpretation is in terms of standard deviation units, making it challenging to interpret the coefficients. However, if the main point of creating a model is to predict observations not in data, or establish probabilities to observations in the data, then interpretations of coefficients may not be important. It depends on why one is modeling the data.

- If the goal of modeling is to understand the relationship between the predictor and response in terms of odds ratios, then care must be taken when transforming individual predictors.
- If the goal of modeling is to assign probabilities to observations in the data, or to calculate probabilities for observations not in the model—but which could be—then optimally transforming predictors is important.

2.6 PREDICTION

2.6.1 Basics of Model Prediction

Prediction is accomplished the same whether we have 1 or greater than 10 predictors in a model. Each predictor is evaluated as adjusted by the other predictors in the model. We discussed the prediction in Chapter 1 in terms of how to calculate the fitted value or probability that $y = 1$ for a logistic model. We will discuss it again later when we evaluate multivariable models and model fit. For now, I shall show how to calculate a predicted probability for a single predictor, how to calculate probabilities for specific predictor values, and then in the next subsection how to construct a confidence interval and graph of a prediction.

```
> summary(logit7 <- glm(died ~ white, family=binomial,
          data=medpar))

Coefficients:
            Estimate Std. Error z value Pr(>|z|)
(Intercept)  -0.9273     0.1969  -4.710 2.48e-06 ***
white         0.3025     0.2049   1.476     0.14
--
    Null deviance: 1922.9 on 1494 degrees of freedom
Residual deviance: 1920.6 on 1493 degrees of freedom
AIC: 1924.6

> exp(coef(logit7))
(Intercept)       white
  0.3956044   1.3532548
```

White patients have a 35% greater odds of death within 48 h of admission than do nonwhite patients.

```
LINEAR PREDICTOR
> etab <- predict(logit7)

FITTED VALUE; PROBABILITY THAT DIED==1
> fitb <- logit7$fitted.value

TABULATION OF PROBABILITIES
> table(fitb)
fitb
0.283464566929547 0.348684210526331
              127              1368
```

1368 *white* patients have an approximate 0.349 probability of dying within 48 h of admission. Nonwhite patients have some 0.283 probability of dying. Since the predictor is binary, there are only two predicted values.

Let us model died on *los,* a continuous predictor.

```
> summary(logit8 <- glm(died ~ los, family=binomial,
          data=medpar))

Coefficients:
             Estimate Std. Error z value Pr(>|z|)
(Intercept) -0.361707   0.088436  -4.090 4.31e-05 ***
los         -0.030483   0.007691  -3.964 7.38e-05 ***
```

```
    Null deviance: 1922.9 on 1494 degrees of freedom
Residual deviance: 1904.6 on 1493 degrees of freedom
AIC: 1908.6

> exp(coef(logit8))
(Intercept)         los
  0.6964864   0.9699768

> etac <- predict(logit8)
> fitc <- logit8$fitted.value
> summary(fitc)
   Min. 1st Qu.  Median    Mean 3rd Qu.    Max.
0.01988 0.31910 0.35310 0.34310 0.38140 0.40320
```

The predicted values of *died* given *los* range from 0.02 to 0.40.

If we wish to determine the probability of death within 48 h of admission for a patient who has stayed in the hospital for 20 days, multiply the coefficient on *los* by 20, add the intercept to obtain the linear predictor for *los* at 20 days. Apply the inverse logit link to obtain the predicted probability.

```
> xb20 <- -0.361707 - 0.030483*20
> mu20 <- 1/(1+exp(-xb20))
> mu20
[1] 0.2746081
```

The probability is 0.275. A patient who stays in the hospital for 20 days has a 27.5% probability of dying within 48 h of admission—given a specific disease from this data.

2.6.2 Prediction Confidence Intervals

We next calculate the standard error of the linear predictor. We use the *predict* function with the `type = "link"` and `se.fit = TRUE` options to place the predictions on the scale of the linear predictor, and to guarantee that the *lpred* object is in fact the standard error of the linear prediction.

```
> lpred <- predict(logit8, newdata=medpar, type="link",
                   se.fit=TRUE)
```

Now we calculate the 95% confidence interval of the linear predictor. As mentioned earlier, we assume that both sides of the distribution are used in determining the confidence interval, which means that 0.025 is taken from each tail of the distribution. In terms of the normal distribution, we see that

```
> up <- lpred$fit+ (qnorm(.975)  * lpred$se.fit)
> lo <- lpred$fit  -  (qnorm(.975)  * lpred$se.fit)
> eta <- lpred$fit
```

We may use the inverse logistic link function, to convert the above three statistics to the probability scale. We could also use the true inverse logit link function, $\exp(xb)/(1 + \exp(xb))$ or $1/(1 + \exp(-xb))$, to convert these to the probability scale. It is easier to simply use the *linkinv* function. A summary of each is displayed based on the following code.

```
> upci <- logit8$family$linkinv(up)
> mu <- logit8$family$linkinv(eta)
> loci <- logit8$family$linkinv(lo)

> summary(loci)
     Min.   1st Qu.   Median      Mean  3rd Qu.      Max.
0.004015 0.293000 0.328700 0.312900 0.350900 0.364900

> summary(mu)
   Min. 1st Qu. Median     Mean 3rd Qu.    Max.
0.01988 0.31910 0.35310 0.34310 0.38140 0.40320

> summary(upci)
   Min. 1st Qu. Median     Mean 3rd Qu.    Max.
0.09265 0.34640 0.37820 0.37540 0.41280 0.44260
```

The mean of the lower 95% confidence interval is 0.313, the mean of μ is 0.343, and the mean of the upper confidence interval is 0.375. A simple R plot of the predicted probability of death for days in the hospital for patients in this data is displayed as (Figure 2.2):

```
> layout(1)
> plot(medpar$los, mu, col=1)
> lines(medpar$los, loci, col=2, type='p')
> lines(medpar$los,upci, col=3, type='p')
```

We next discuss logistic models with more than one predictor. These are the types of models that are in fact employed in real-life studies and projects. Understanding single predictor models, however, provides a solid basis for understanding more complex models.

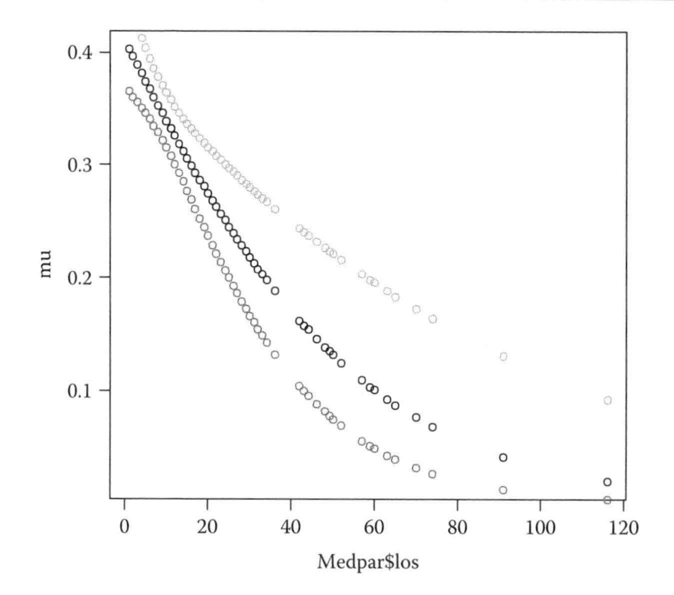

FIGURE 2.2 Probability of length of stay by type of admission.

SAS CODE

```
/* Section 2.1 */

*Create a new dataset with binary variables x and y;
data xdta;
        input x y @@;
        datalines;
 1 1 0 0 1 0 0 1 1
 0 1 1 1 0 0 1 0 1
 ;
 run;

*Build the logistic model;
proc genmod data=xdta descending;
        model y=x / dist=binomial link=logit;
        output out=residual resdev=deviance;
run;

*Another way to build the logistic model;
proc logistic data=xdta descending;
        model y=x / clparm=both;
        output out=residual resdev=deviance;
run;
```

```
*Statistics of deviance residual;
proc means data=residual min q1 median q3 max maxdec=4;
        var deviance;
run;

*Generate a table of y by x;
proc freq data=xdta;
        tables y*x / norow nocol nocum nopercent;
run;

*Expb option provides the odds ratio;
proc genmod data=xdta descending;
        model y=x / dist=binomial link=logit;
        estimate "Intercept" Intercept 1 / exp;
        estimate "x" x 1 / exp;
run;

/* Section 2.2 */

*Refer to proc genmod in section 2.1 to build the logistic model;

*Create a dataset to make calculations;
data data1;
        set xdta;
        if x=1 then xb=1.0986-1.5041*1;
        else if x=0 then xb=1.0986-1.5041*0;
        mu=1/(1+exp(-xb));
        o=mu/(1-mu);
        or=o;
        if or < 1 then or=0.667/3;
        coeff=log(or);
        format mu 4.2 o or xb coeff 7.4;
run;

*Print the dataset;
proc print data=data1;
        var x o;
run;

*Print the whole dataset;
proc print data=data1;
run;

/* Section 2.3 */

*Build the logistic model- covb option provides var-cov matrix;
proc genmod data=xdta descending;
        model y=x / dist=binomial  link=logit covb;
run;

*Use SAS interactive matrix language;
proc iml;
        vcov={1.33333 -1.33333,
             -1.33333  2.16667};
```

```
        se=sqrt(diag(vcov));
        print se;
quit;

*Logistic regression with OIM standard error;
proc surveylogistic data=xdta;
        model y(event='1')=x;
run;

*Refer to proc genmod in section 2.3 to obtain var-cov matrix;

*Calculations of odds ratio and model statistics;
proc iml;
        vcov={1.33333 -1.33333,
             -1.33333  2.16667};
        coef={1.0986, -1.5041};
        or=exp(coef);
        se=sqrt(diag(vcov));
        ose=se*or;
        print or [format = 7.4] ose [format = 7.4];

        zscore=coef/se;
        delta=ose;
        z=zscore[,+];
        pvalue=2*(1-probnorm((abs(z))));
        print z pvalue;

        se1=se[,+];
        loci=coef-quantile('normal', 0.975)*se1;
        upci=coef+quantile('normal', 0.975)*se1;
        expl=exp(loci);
        expu=exp(upci);
        print or [format=7.4] delta [format=7.4] z [format=7.4]
            pvalue [format=7.4] expl [format=7.4] expu [format=7.4];
quit;

*Clparm=both provides both PL and Wald confidence intervals;
proc logistic data=xdta descending;
        model y=x / clparm=both;
run;

/* Section 2.4 */

*Refer to the code in section 1.4 to import and print medpar dataset;

*Generate the frequency table of type and output the dataset;
proc freq data=medpar;
        tables type / out=freq;
run;

*Build the logistic model with class;
proc genmod data=medpar descending;
        class type (ref='1') / param = ref;
        model died=type / dist=binomial link=logit;
```

```
        estimate "Intercept" Intercept 1 / exp;
        estimate "type2" type 1 0 / exp;
        estimate "type3" type 0 1 / exp;
        output out=residual resdev=deviance;
run;

*Set up format for variable type;
proc format;
        value typefmt 1="Elective Admit"
                      2="Urgent Admit"
                      3="Emergency Admit";
run;

*Logistic regression with controlled reference;
proc genmod data=medpar descending;
        class type (ref='Elective Admit') / param = ref;
        model died=type / dist=binomial link=logit;
        estimate "Intercept" Intercept 1 / exp;
        estimate "type2" type 1 0 / exp;
        estimate "type3" type 0 1 / exp;
        format type typefmt.;
run;

*Logistic regression with controlled reference;
proc genmod data=medpar descending;
        class type (ref='Emergency Admit') / param = ref;
        model died=type / dist=binomial link=logit;
        estimate "Intercept" Intercept 1 / exp;
        estimate "type2" type 1 0 / exp;
        estimate "type3" type 0 1 / exp;
        format type typefmt.;
run;

*Refer to proc freq in section 2.4 to generate the frequency table;

*Re-categorized variable type;
data medpar1;
        set medpar;
        if type in (2,3) then type=2;
run;

*Refer to proc freq in section 2.4 to generate the frequency table;

*Logistic regression with re-categorized type;
proc genmod data=medpar1 descending;
        class type (ref='1') / param = ref;
        model died=type / dist=binomial link=logit;
        estimate "Intercept" Intercept 1 / exp;
        estimate "type2" type 1 0 / exp;
        estimate "type3" type 0 1 / exp;
run;
```

```
/* Section 2.5 */

*Summary for variable los;
proc means data=medpar min q1 median mean q3 max maxdec=3;
        var los;
run;

*Build the generalized additive model;
proc gam data=medpar;
        model died (event='1')=spline(los) / dist=binomial;
run;

*Refer to the code in section 1.4 to import and print badhealth
dataset;

*Refer to proc freq in section 2.4 to generate the frequency table;

*Summary for variable age;
proc means data=badhealth min q1 median mean q3 max maxdec=2;
        var age;
        output out=center mean=;
run;

*Create a macro variable;
proc sql;
        select age into: mean
        from center;
quit;

*Refer to proc means in section 2.5 to summarize numvisit;

*Center the age;
data badhealth1;
        set badhealth;
        cage=age-&mean;
run;

*Refer to proc means in section 2.5 to summarize centered age;

*Provide the std;
proc means data=badhealth std;
        var age;
        output out=stderror std=;
run;

*Create a macro variable;
proc sql;
        select age into: std
        from stderror;
quit;

*Scale age with a different way;
proc standard data=badhealth mean=0 std=&std out=cenage;
var age;
run;
```

```
*Build the logistic model;
proc genmod data=badhealth descending;
        model badh=age / dist=binomial link=logit;
run;

*Build the logistic model with centered age;
proc genmod data=badhealth1 descending ;
        model badh=cage / dist=binomial  link=logit;
run;

*Standardize age and output the sage dataset;
proc standard data=badhealth mean=0 std=1 out=sage;
        var age;
run;

*Build the logistic model with standardized age;
proc genmod data=sage descending ;
        model badh=age / dist=binomial  link=logit;
run;

/* Section 2.6 */

*Build the logistic model and output model prediction;
proc genmod data=medpar descending;
        model died=white / dist=binomial link=logit;
        output out=etab pred=fitb;
run;

*Refer to proc freq in section 2.4 to generate the frequency table;

*Build the logistic model and output model prediction;
proc genmod data=medpar descending;
        model died=white / dist=binomial link=logit;
        output out=etac pred=fitc;
run;

*Refer to proc means in section 2.5 to summarize fitc;

*Create a dataset to make calculations;
data prob;
        xb20=-0.3617 - 0.0305*20;
        mu20=1/(1+exp(-xb20));
run;

*Print the variable mu20;
proc print data=prob;
        var mu20;
run;

*Build the logistic model and output confidence intervals;
proc genmod data=medpar descending;
        model died=los / dist=binomial  link=logit;
        output out=cl pred=mu lower=loci upper=upci;
run;
```

```
*Summary for confidence intervals;
proc means data=cl min q1 median mean q3 max maxdec=5;
        var loci mu upci;
run;

*Graph scatter plot;
proc sgplot data=cl;
        scatter x=los y=mu;
        scatter x=los y=loci;
        scatter x=los y=upci;
run;
```

STATA CODE

```
2.1
. use xdta
. list
. glm y x, fam(bin) nolog
. table y x
. tab y x
. glm y x, fam(bin) eform nolog nohead

2.2
. glm y x, fam(bin) nolog nohead
. di 1.098612 - 1.504077*1
. di 1.098612 - 1.504077*0
. predict xb, xb
. predict mu
. gen o=mu/(1-mu)
. gen or=.6666667/3 if o<1
. replace or=o if or==.
. gen coef=log(or)
. l y x xb mu o or coef

2.3
. glm y x, fam(bin) nolog nohead
. estat vce
. glm y x, fam(bin) nolog nohead scale(x2)
. glm y x, fam(bin) nolog nohead eform
. di normal(-abs(_b[x]/_se[x]))*2              // p-value for x
. di normal(-abs(_b[_cons]/_se[_cons]))*2     // p-value for intercept
. use medpar, clear
. glm died white los i.type, fam(bin) nolog nohead
. glm died white los i.type, fam(bin) nolog nohead eform

2.4
. use medpar, clear
. list in 1/6
. tab type
```

```
. glm died i.type, fam(bin) nolog nohead
. glm died i.type, fam(bin) nolog nohead eform
. glm died b3.type, fam(bin) nolog nohead
. tab type, gen(type)
. gen type23 = type2 | type3
. tab type23
```

2.5

```
. use badhealth, clear
. list in 1/6
. tab badh
. summary age
. summary numvis
. egen meanage = mean(age)
. gen cage = age - meanage
. * or: center age, pre(c)
. glm badh cage, fam(bin) nolog nohead
. center age, pre(s) stand
. glm badh sage, fam(bin) nolog nohead
```

2.6

```
. glm died white, fam(bin) nolog nohead
. glm died white, fam(bin) nolog nohead eform
. predict etab, xb
. predict fitb, mu
. tab fitb
. glm died los, fam(bin)
. glm died los, fam(bin) eform
. predict etac, xb
. predict fitc
. summary fitc
. use medpar
. glm died los, family(bin) nolog
. predict eta, xb                  // linear predictor; eta
. predict se_eta, stdp             // standard error of the prediction
. gen mu = exp(eta)/(1 + exp(eta))   // or: predict mu
. gen low = eta - invnormal(0.975) * se_eta
. gen up = eta + invnormal(0.975) * se_eta
. gen lci = exp(low)/(1 + exp(low))
. gen uci = exp(up)/(1 + exp(up))
. sum lci mu uci
. scatter mu lci uci los
```

Logistic Models *Multiple Predictors* 3

3.1 SELECTION AND INTERPRETATION OF PREDICTORS

The logic of modeling data with logistic regression changes very little when more predictors are added to a model. The basic logistic regression formula we displayed becomes more meaningful when there is more than one predictor in a model. Equation 3.1 below expresses the relationship of each predictor to the predicted linear predictor, $x_i'\beta$, or η_i. It is more accurate to symbolize the predicted linear predictor as $\hat{\eta}_i$ or as $x_i'\beta$, but we shall not employ the *hat* symbol on η or β for ease of interpretation, as we have done so for the predicted probability, $\hat{\mu}$. We shall remember from the context that the expression is predicted or estimated, and not simply given as raw data.

$$\ln\left(\frac{\mu_i}{1-\mu_i}\right) = \eta_i = x_i b = \sum_{i=1}^{n} \beta_0 + x_{i1}'\beta_1 + x_{i2}'\beta_2 + \cdots + x_{ij}'\beta_j \tag{3.1}$$

With respect to logistic regression, each β in Equation 3.1 above indicates a separate coefficient, or slope. Each is interpreted as a partial derivative in calculus. When a predictor is being interpreted, it is in terms of its associated coefficient or rate of change with respect to the response. Each coefficient assumes that when it is interpreted, the other predictors are held as constant.

Each term, or $x'\beta$, in the regression equation indicates that for a one-unit change in the predictor, x, the log-odds of the response changes by β, given that the other terms in the model are held constant. When the logistic regression term is exponentiated, interpretation is given in terms of an odds ratio, rather than log-odds. We can see this in Equation 3.2 below, which results by exponentiating each side of Equation 3.1.

$$\frac{\mu_i}{1-\mu_i} = e^{\sum_{i=1}^{n} \beta_0 + x'_{i1}\beta_1 + x'_{i2}\beta_2 + \cdots + x'_{ij}\beta_j} \tag{3.2}$$

or

$$\frac{\mu_i}{1-\mu_i} = \sum_{i=1}^{n} \exp(\beta_0) + \exp(x'_{i1}\beta_1) + \exp(x'_{i2}\beta_2) + \cdots + \exp(x'_{ij}\beta_j) \tag{3.3}$$

An example will help clarify what is meant when interpreting a logistic regression model. Let's use data from the social sciences regarding the relationship of whether a person identifies themselves as religious. Our main interest will be in assessing how level of education affects religiosity. We'll also adjust by gender (*male*), *age*, and whether the person in the study has children (*kids*). There are 601 subjects in the study, so there is no concern about sample size. The data are in the *edrelig* data set.

A study subject's level of education is a categorical variable with three fairly equal-sized levels: AA, BA, and MA/PhD. All subjects have achieved at least an associate's degree at a 2-year institution. A tabulation of the *educlevel* predictor is shown below, together with the top six values of all variables in the data.

```
> load("c://rfiles/edrelig.rdata")
> head(edrelig)
  male age kids educlevel religious
1    1  37    0    MA/PhD         0
2    0  27    0        AA         1
3    1  27    0    MA/PhD         0
4    0  32    1        AA         0
5    0  27    1        BA         0
6    1  57    1    MA/PhD         1

> table(edrelig$educlevel)

   AA     BA MA/PhD
  205    204    192
```

Male and *kids* are both binary predictors, having values of 0 and 1. 1 indicates (most always) that the name of the predictor is the case. For instance, the binary predictor *male* is 1 = *male* and 0 = *female*. *Kids* = 1 if the subject has children, and 0 if they have no children. *Age* is a categorical variable with levels as 5-year age groups. The range is from 17 to 57. I will interpret *age*, however, as a continuous predictor, with each ascending age as a 5-year period.

We model the data as before, but simply add more predictors in the model. The categorical *educlevel* predictor is factored into its three levels, with the lowest level, AA, as the reference. It is not displayed in model output.

```
> summary(ed1 <- glm(religious ~ age + male + kids + factor(educlevel),
+   family=binomial, data=edrelig))

Coefficients:
                         Estimate Std. Error z value Pr(>|z|)
(Intercept)              -1.43522    0.32996  -4.350 1.36e-05 ***
age                       0.03983    0.01036   3.845 0.000121 ***
male                      0.18997    0.18572   1.023 0.306381
kids                      0.12393    0.21037   0.589 0.555790
factor(educlevel)BA      -0.47231    0.20822  -2.268 0.023313 *
factor(educlevel)MA/PhD  -0.49543    0.22621  -2.190 0.028513 *
---
    Null deviance: 822.21 on 600 degrees of freedom
Residual deviance: 792.84 on 595 degrees of freedom
AIC: 804.84
```

The odds ratios are obtained by:

```
> or <- exp(coef(ed1))
> round(or,4)
            (Intercept)                     age                    male
                 0.2381                  1.0406                  1.2092
                   kids     factor(educlevel)BA factor(educlevel)MA/PhD
                 1.1319                  0.6236                  0.6093
```

Or we can view the entire table of odds ratio estimates and associated statistics using the code developed in the previous chapter.

```
> coef <- ed1$coef
> se <- sqrt(diag(vcov(ed1)))
> zscore <- coef / se
> or <- exp(coef)
> delta <- or * se
> pvalue <- 2*pnorm(abs(zscore),lower.tail=FALSE)
> loci <- coef - qnorm(.975) * se
> upci <- coef+qnorm(.975) * se
> ortab <- data.frame(or, delta, zscore, pvalue, exp(loci), exp(upci))
> round(ortab, 4)
                            or  delta zscore pvalue  exp.loci. exp.upci.
(Intercept)             0.2381 0.0786 -4.3497 0.0000    0.1247    0.4545
age                     1.0406 0.0108  3.8449 0.0001    1.0197    1.0620
male                    1.2092 0.2246  1.0228 0.3064    0.8403    1.7402
kids                    1.1319 0.2381  0.5891 0.5558    0.7495    1.7096
factor(educlevel)BA     0.6236 0.1298 -2.2683 0.0233    0.4146    0.9378
factor(educlevel)MA/PhD 0.6093 0.1378 -2.1902 0.0285    0.3911    0.9493
```

Since we are including more than a single predictor in this model, it's wise to check additional model statistics. Interpretation gives us the following, with the understanding that the values of the other predictors in the model are held constant.

age	Subjects in a higher 5-year age group have a 4% greater odds of being religious than those in the lower age division, assuming that the values of other predictors are constant (at their mean).
male	Males have a some 21% greater odds of being religious than females.
kids	Study subjects having children have a 13% higher odds of being religious than are those without children.
educlevel	Those in the study whose highest degree is a BA have a 60% greater odds of being nonreligious compared to those whose highest degree is an AA.

```
> 1/.6235619
[1] 1.60369
```

Those in the study whose highest degree is a MA/PhD have a 64% greater odds of being nonreligious compared to those whose highest degree is an AA.

```
> 1/.6093109
[1] 1.641198
```

Notice that I switched the interpretation of the levels of *educlevel* so that they are positive. I could have said that those with a BA have 40% less odds of being religious than are those with a highest degree of AA. The interpretation is not going to be understood as well as if I express a positive relationship. I recommend employing a positive interpretation of odds ratio if it makes sense in the context of the study.

3.2 STATISTICS IN A LOGISTIC MODEL

When logistic regression is used from within GLM software, the output is pretty much the same regardless of the software package used for the modeling process. R is an exception though. The logic of basic R is to display the minimum for the statistical procedure being executed, but provide options by which the user can display additional statistics. The R *summary* function is such an option, providing substantially more model statistics than are provided by simply using the *glm* function when modeling a logistic regression. In this section, we define and discuss the various GLM statistics that are provided in the R's summary function. Stata, SAS, SPSS, Limdep, and other GLM software generally provide the same statistics.

I shall display the *ed1* model we just estimated, removing the coefficient table from our view. We are only interested here in the ancillary model statistics that can be used for evaluating model fit.

R

```
> summary(ed1 <- glm(religious ~ age + male + kids+ factor(educlevel),
+         family = binomial, data = edrelig))

Deviance Residuals:
    Min        1Q    Median        3Q       Max
-1.6877   -1.0359   -0.8467    1.2388    1.6452
                                  . . .
    Null deviance: 822.21   on 600   degrees of freedom
Residual deviance: 792.84   on 595   degrees of freedom
AIC: 804.84
```

R provides a summary table of deviance statistics, together with null and residual deviance statistics and their respective degrees of freedom. Stata also has a deviance statistic, and the corresponding degrees of freedom. These were thought to be very important statistics needed for fit analysis, but their important has waned in recent times.

The deviance statistic is based on the log-likelihood. Keep in mind that many fit statistics are based on the log-likelihood function. Stata displays the log-likelihood together with the table of estimates. R's *glm* function does not have the log-likelihood saved post estimation statistic.

Recall that the Bernoulli distribution log-likelihood can be given in exponential family form as

$$\mathcal{L}(\mu_i; y_i) = \sum_{i=1}^{n} y_i \ln\left(\frac{\mu_i}{1-\mu_i}\right) + \ln(1-\mu_i) \tag{3.4}$$

The deviance is calculated as a goodness-of-fit test for GLM models, and is defined as twice the difference between the saturated log-likelihood minus the full log-likelihood. The saturated model has a parameter for each observation in the model. This means that a y replaces every μ in the log-likelihood function.

$$D = 2\sum_{i=1}^{n}\{\mathcal{L}(y_i;y_i) - \mathcal{L}(\mu_i;y_i)\} \tag{3.5}$$

The Bernoulli deviance is expressed as:

Logistic Model Deviance Statistic:

$$D = 2\sum_{i=1}^{n} y_i \ln\left(\frac{y_i}{\mu_i}\right) + (1 - y_i)\ln\left(\frac{1 - y_i}{1 - \mu_i}\right) \tag{3.6}$$

For GLM models, including logistic regression, the deviance statistic is the basis of model convergence. Since the GLM estimating algorithm is iterative, convergence is achieved when the difference between two deviance values reaches a threshold criterion, usually set at 0.000001.

When convergence is achieved, the values of the coefficients, of mu, eta, and other statistics are at their optimal values.

The other main use of the deviance statistic is as a goodness-of-fit test. The "residual" deviance is the value of D that can be calculated following model convergence. Each observation will have a calculated value of D_i as $y*\ln(y/\mu) + (1 - y)*\ln[(1 - y)/(1 - \mu)]$. Sum the Ds across all observations and multiply by 2—that's the deviance statistic. The value of the deviance statistic for an intercept only model; that is, a model with no predictors, is called the *null deviance.* The null degrees of freedom is the total observations in the model minus the intercept, or $n - 1$. The residual degrees of freedom is n minus the number of predictors in the model, including the intercept. For the example model, there are 601 observations and six predictors: *age*, *male*, *kids*, *educlevel(BA)*, *educlevel(MA/PhD)*, and the *intercept*. The reference level is not counted. The null deviance degrees of freedom (*dof*) is 600 and residual *dof* is 595. A traditional fit statistic we shall discuss in the next chapter is based on the *Chi*2 distribution of the deviance with a *dof* of the residual deviance.

The Pearson *Chi*2 goodness-of-fit statistic is defined as the square of the raw residual divided by the variance statistic. The raw residual is the value of the response, y, minus the mean (μ). The Bernoulli variance function for the logistic regression model is $\mu(1 - \mu)$. Therefore,

Pearson Chi2 GOF Statistic:

$$\sum_{i=1}^{n}\frac{(y_i - \mu_i)^2}{V(\mu_i)} \tag{3.7}$$

Logistic Model Pearson Chi2 GOF Statistic (based on the Bernoulli distribution):

$$\sum_{i=1}^{n} \frac{(y_i - \mu_i)^2}{\mu_i(1 - \mu_i)} \tag{3.8}$$

The degrees of freedom for the Pearson statistic are the same as for the deviance. For count models, the dispersion statistic is defined as the Pearson *Chi*2 statistic divided by the residual *dof*. Values greater than 1 indicate possible overdispersion. The same is the case with grouped logistic models—a topic we shall discuss in Chapter 5. The deviance dispersion can also be used for binomial models—again a subject to which we shall later return.

I mentioned earlier that raw residuals are defined as "$y - \mu$." All other residuals are adjustments to this basic residual. The Pearson residual, for example, is defined as:

Pearson Residual:

$$\frac{y - \mu}{\sqrt{\mu(1 - \mu)}} \tag{3.9}$$

It is important to know that the sum of the squared Pearson residuals is the Pearson *Chi*2 statistic:

$$\text{Pearson } Chi2 \text{ statistic} = \sum_{i=1}^{n} \left(\frac{y_i - \mu_i}{\sqrt{\mu_i(1 - \mu_i)}} \right)^2 \tag{3.10}$$

In fact, the way programmers calculate the Pearson *Chi*2 statistic is by summing the squared Pearson residuals.

```
> pr <- resid(ed1, type = "pearson") # calculates Pearson residuals
> pchi2 <- sum(residuals(ed1, type = "pearson")^2)
> disp <- pchi2/ed1$df.residual
> c(pchi2, disp)
[1] 600.179386 1.008705
```

Unfortunately neither the Pearson *Chi*2 statistic nor the Pearson dispersion is directly available from R. Strangely though, the Pearson dispersion is used to generate what are called *quasibinomial* models; that is, logistic models with too much or too little correlation in the data. See Hilbe (2009) and Hilbe and Robinson (2013) for a detailed discussion of this topic.

I created a function that calculates the Pearson *Chi2* and dispersion following *glm* estimation. Called *P__disp*, it is a function in the COUNT and LOGIT packages. If the name of the model of concern is *mymodel*, type *P__disp(mymodel)* on the command line.

Deviance residuals are calculated on the basis of the deviance statistic defined above. For binary logistic regression, deviance residuals take the form of

If $y = 1$,

$$\text{sign}(y - \mu) * \sqrt{2\sum \ln\left(\frac{1}{\mu}\right)} \tag{3.11}$$

If $y = 0$,

$$\text{sign}(y - \mu) * \sqrt{2\sum \ln\left(\frac{1}{1-\mu}\right)} \tag{3.12}$$

Using R's built-in deviance residual option for *glm* models, we may calculate a summary of the values as,

```
> dr <-resid(ed1, type="deviance")
> round(summary(dr),4)
   Min. 1st Qu.  Median    Mean 3rd Qu.    Max.
-1.6880 -1.0360 -0.8467 -0.0434  1.2390  1.6450
```

Note the closeness of the residual values to what is displayed in the *ed1* model output at the beginning of this section.

I should also mention that the above output for Pearson residuals informs us that the dispersion parameter for the model is 1 (1.008705). The logistic model is based on the Bernoulli distribution with only a mean parameter. There is no scale parameter for the Bernoulli distribution. The same is the case for the Poisson count model. In such a case the software reports that the value is 1, which means that it cannot affect the other model statistics or the mean parameter. It is statistically preferred to use the term scale in this context than it is dispersion, for reasons that go beyond this text. See Hilbe (2011) or Hilbe (2014) for details.

The *glm* function fails to display or save the log-likelihood function, although it is used in the calculation of other saved statistics. By back-coding other statistics an analyst can calculate a statistic such as the log-likelihood which is given at the start of this section. For the *ed1* model,

Log-likelihood:

```
> (ed1$df.null - ed1$df.residual+1) - ed1$aic/2
[1] -396.4198
```

Other important statistics which will be required when we set up residual analysis are the *hat* matrix diagonal and standardized Pearson and standardized deviance residuals. These may be calculated for the *ed1* model using *ht* code.

The hat matrix diagonal is defined as:

$$h = hat = W^{1/2}X(X'WX)^{-1}X'W^{1/2} \tag{3.13}$$

with W as a weight defined as diag{1/(μ(1 − μ))*ln(μ/(1 − μ))²} and X as the predictor matrix. The hat statistic can also be calculated as the Bernoulli variance times the square of the standard error of prediction. R does not have a precalculated function for the standard error of the predicted value, but several other statistical packages do; for example, Stata, SAS. The R *glm* function does have, however, a function to calculate *hat* values, as we can observe below. Note that standardization of the Pearson and deviance residuals is accomplished by dividing them each by the square root of 1 − *hat*.

Hat Matrix Diagonal Influence Statistic:

```
> hat <- hatvalues(ed1)
> summary(hat)
    Min.  1st Qu.   Median     Mean  3rd Qu.     Max.
0.006333 0.007721 0.009291 0.009983 0.011220 0.037910
```

Standardize Pearson Residual:

```
> stpr <- pr/sqrt(1-hat)
> summary(stpr)
     Min.   1st Qu.    Median      Mean   3rd Qu.      Max.
-1.791000 -0.845700 -0.660300 -0.002086 1.078000 1.705000
```

Standardized Deviance Residual:

```
> stdr <- dr/sqrt(1-hat)
> summary(stdr)
    Min.  1st Qu.   Median     Mean  3rd Qu.     Max.
-1.70200 -1.04000 -0.85150 -0.04356 1.24300   1.65600
```

R's *glm* function saves other statistics as well. To identify statistics that can be used following an R function, type ?glm or ? followed by the function name, and a help file will appear with information about model use and saved values. All CRAN functions should have a help document associated with the function, but packages and functions that are not part of the CRAN family have no such requirements. Nearly all Stata commands or functions have associated help. For general help on *glm* type, "help glm."

3.3 INFORMATION CRITERION TESTS

Information criterion tests are single statistics by which analysts may compare models. Models with lower values of the same information criterion are considered better fitted models. A number of information tests have been published, but only a few are frequently used in research reports.

3.3.1 Akaike Information Criterion

The Akaike information criterion (AIC) test, named after Japanese statistician Hirotsugu Akaike (1927–2009), is perhaps the most well-known and well used information statistic in current research. What may seem surprising to many readers is that there are a plethora of journal articles detailing studies proving how poor the AIC test is in assessing which of two models is the better fitted. Even Akaike himself later developed another criterion which he preferred to the original. However, it is his original 1973 version that is used by most researchers and that is found in most journals to assess comparative model fit.

The traditional AIC statistic is found in two versions:

$$\text{AIC} = -2\mathcal{L} + 2k \text{ or } -2(\mathcal{L} - k) \tag{3.14}$$

or

$$\text{AIC} = \frac{-2\mathcal{L} + 2k}{n} \text{ or } \frac{-2(\mathcal{L} - k)}{n} \tag{3.15}$$

where $\mathcal{L}$ is the model log-likelihood, k is the number of parameter estimates in the model, and n is the number of observations in the model. For logistic regression, parameter estimates are the same as predictors, including the intercept. Using the *medpar* data set described earlier, we model *died* on

```
> data(medpar)
> summary(mymod <- glm(died ~ white + hmo + los + factor(type),
+                      family=binomial,
+                      data=medpar))

Coefficients:
             Estimate Std. Error z value Pr(>|z|)
(Intercept) -0.720149   0.219073  -3.287  0.00101 **
white        0.303663   0.209120   1.452  0.14647
```

```
hmo              0.027204   0.151242   0.180  0.85725
los             -0.037193   0.007799  -4.769 1.85e-06 ***
factor(type)2   0.417873   0.144318   2.896  0.00379 **
factor(type)3   0.933819   0.229412   4.070 4.69e-05 ***

    Null deviance: 1922.9 on 1494 degrees of freedom
Residual deviance: 1881.2 on 1489 degrees of freedom
AIC: 1893.2
```

Using R and the values of the log-likelihood and the number of predictors, we may calculate the AIC as:

```
> -2*( -940.5755 -6)
[1] 1893.151
```

This is the same value that is displayed in the *glm* output. It should be noted that of all the information criteria that have been formulated, this version of the AIC is the only one that does not adjust the log-likelihood by n, the number of observations in the model. All others adjust by some variation of number of predictors and observations. If the AIC is used to compare models, where n is different (which normally should not be the case), then the test will be mistaken. Using the version of AIC where the statistic is divided by n is then preferable—and similar to that of other criterion tests. The AIC statistic, captured from the postestimation statistics following the execution of *glm*, is displayed below, as is AICn. These statistics are also part of the *modelfit* function described below.

```
AIC - from postestimation statistics
> mymod$aic
[1] 1893.151

AICn - AIC divided by n
> aicn <- mymod$aic/(mymod$df.null+1)
> aicn
[1] 1.266322
```

3.3.2 Finite Sample

Finite sample AIC was designed to compare logistic models. It is rarely used in reports, but is important to know. It is defined as:

$$\text{FAIC} = \frac{-2\{[\mathcal{L} - k - k(k+1)]/(n-k-1)\}}{n} \tag{3.16}$$

3.3.3 Bayesian Information Criterion

The Schwarz Bayesian information criterion (BIC) is the most used BIC test found in the literature. Developed by Gideon Schwarz in 1978, its value differs little from the AIC statistic. Most statisticians prefer the use of this statistic, but the AIC nevertheless appears to be more popular. My recommendation is to test models with both statistics. If the values substantially differ, it is likely that the model is mis-specified. Another binomial link may be required; for example, probit, complementary loglog, or loglog.

$$\text{BIC} = -2\mathcal{L} + k\ln(n) \tag{3.17}$$

The BIC is not available in the default R software, but it can be obtained by using the *modelfit* function in the COUNT package. Following the *mymod* model above, an analyst should type

```
> modelfit(mymod)
$AIC
[1] 1893.151

$AICn
[1] 1.266322

$BIC
[1] 1925.01

$BICqh
[1] 1.272677
```

The BIC statistic is given as 1925.01, which is the same value displayed in the Stata *estat ic* post-estimation command. Keep in mind that size comparison between the AIC and BIC statistics are not statistically meaningful.

3.3.4 Other Information Criterion Tests

The Hannan and Quinn BIC statistic, first developed in 1979, is provided in the *modelfit* output. It has a value similar to AIC/n. If the two statistics differ by much, this is an indication of mis-specification.

Another test statistic developed for correlated data is called the AIC_H statistic. See Hilbe (2014) for a full discussion:

$$\text{AIC}_\text{H} = -2\mathcal{L} + \frac{4(p^2 - pk - 2p)(p + k + 1)(p + k + 2)}{n - p - k - 2} \tag{3.18}$$

3.4 THE MODEL FITTING PROCESS: ADJUSTING STANDARD ERRORS

When dealing with logistic models we must be concerned, among other things, with data that may be more correlated than is allowed by the underlying Bernoulli distribution. If the data are taken from clusters of items; for example, litters of pups, galaxies, schools within a city, and so forth, the independence of observations criterion of the Bernoulli PDF and likelihood is violated. Standard errors based on this distribution will be biased, indicating that a predictor significantly contributes to a model when it in fact does not.

Other distributional problems can exist as well. We need not describe them all here. Just be aware that it is wise to check for problems that the model has with the data. Fortunately, we do not have to identify every problem that may exist in the data in order to produce a well-fitted model. At times performing various adjustments to the model at the outset will allow the analyst to model the data without having to be concerned about the particular cause.

3.4.1 Scaling Standard Errors

I described why we may need to scale standard errors in Chapter 2, Section 2.3.1, and also gave an example of how to do it. To repeat, if the data are correlated or somehow do not meet the distribution requirements of the PDF underlying the model, the standard errors displayed in model results are likely biased. When we scale standard errors, we are adjusting them to the values they would have if there was not extra correlation or some other problem with the data. Scaling was designed to deal with excess correlation in the data, but it also can be used to address other unknown problems.

R users may scale the standard errors of a logistic model by using the quasibinomial "family" with the *glm* function. Scaling is accomplished by multiplying the square root of the Pearson dispersion by the standard error of the model. When we discussed scaling in Chapter 2, we had not yet discussed the dispersion statistic, which is essential the operation.

Let's use the *mymod* example we have been using in this chapter to show how to scale standard errors. By creating them by hand, it will allow us to better understand what they are doing. First, let's show a table with the model coefficients and model standard errors

```
> coef <- mymod$coefficients
> se <- sqrt(diag(vcov(mymod)))
```

```
> coefse <- data.frame(coef, se)
> coefse
                       coef          se
(Intercept)     -0.72014852 0.21907288
white            0.30366254 0.20912002
hmo              0.02720413 0.15124225
los             -0.03719338 0.00779851
factor(type)2    0.41787319 0.14431763
factor(type)3    0.93381912 0.22941205
```

Next we create Pearson dispersion statistics and multiply their square root by *se* above.

```
> pr <- resid(mymod, type="pearson")
> pchi2 <- sum(residuals(mymod, type="pearson")^2)
> disp <- pchi2/mymod$df.residual
> scse <- se*sqrt(disp)
> newcoefse <- data.frame( coef, se, scse)
> newcoefse
                       coef          se        scse
(Intercept)     -0.72014852 0.21907288 0.221301687
white            0.30366254 0.20912002 0.211247566
hmo              0.02720413 0.15124225 0.152780959
los             -0.03719338 0.00779851 0.007877851
factor(type)2    0.41787319 0.14431763 0.145785892
factor(type)3    0.93381912 0.22941205 0.231746042
```

We can now check to see if the *quasibinomial* "family" option produces scaled standard errors

```
> summary(qmymod <- glm(died ~ white + hmo + los + factor(type),
+                   family = quasibinomial,
+                   data = medpar))

                              . . .

Coefficients:
                Estimate Std. Error t value Pr(>|t|)
(Intercept)    -0.720149   0.221302  -3.254  0.00116  **
white           0.303663   0.211248   1.437  0.15079
hmo             0.027204   0.152781   0.178  0.85870
los            -0.037193   0.007878  -4.721  2.56e-06 ***
factor(type)2   0.417873   0.145786   2.866  0.00421  **
factor(type)3   0.933819   0.231746   4.029  5.87e-05 ***
```

```
---(Dispersion parameter for quasibinomial family
taken to be 1.020452)

    Null deviance: 1922.9 on 1494 degrees of freedom
Residual deviance: 1881.2 on 1489 degrees of freedom
AIC: NA
```

The standard errors displayed in the quasibinomial model are identical to the scaled standard errors we created by hand. Remember that there is no true *quasibinomial* GLM family. *Quasibinomial* is not a separate PDF. It is simply an operation to provide scaled standard errors on a binomial model such as logistic regression.

When an analyst models a logistic regression with scaled standard errors, the resultant standard errors will be identical to model-based standard errors if there are no distributional problems with the data. In other words, a logistic model is not adversely affected if standard errors are scaled when they do not need it.

A caveat when using R's *quasibinomial* family: *p*-values are based on *t* and not *z* as they should be. As a result a predictor *p*-value may be >0.05 and its confidence interval not include 0. Our *toOR* function used with quasibinomial models provides correct values. To see this occur, model the grouped *quasibinomial* model: sick <- c(77,19,47,48,16,31); cases <- c(458,147,494,384, 127,464); feed <- c(1,2,3,1,2,3); gender <- c(1,1,1,0,0,0).

3.4.2 Robust or Sandwich Variance Estimators

Scaling was the foremost method of adjusting standard errors for many years—until analysts began to use what are called robust or sandwich standard errors. Like scaling, using robust standard errors only affects the model when there are problems with the model-based standard errors. If there is none, then the robust standard error reduces to the model-based errors. Many statisticians recommend that robust or sandwich standard errors be used as a default.

I shall use the same data to model a logistic regression with sandwich or robust standard errors. The sandwich package must be installed and loaded before being able to create sandwich standard errors.

```
>library(sandwich)
> rmymod <- glm(died ~ white + hmo + los + factor(type),
                 family = binomial, data = medpar)
> rse <- sqrt(diag(vcovHC(rmymod, type = "HC0")))
```

The robust standard errors are stored in *rse*. We'll add those to the table of standard errors we have been expanding.

```
> newcoefse2 <- data.frame( coef, se, scse, rse)
> newcoefse2
                      coef         se        scse         rse
(Intercept)    -0.72014852 0.21907288 0.221301687 0.219434958
white           0.30366254 0.20912002 0.211247566 0.210398430
hmo             0.02720413 0.15124225 0.152780959 0.150972915
los            -0.03719338 0.00779851 0.007877851 0.009726677
factor(type)2   0.41787319 0.14431763 0.145785892 0.145242836
factor(type)3   0.93381912 0.22941205 0.231746042 0.229306158
```

It should be mentioned that models evaluating longitudinal and clustered data like Generalized Estimating Equations always assume that there is more correlation within longitudinal units or panels, or within cluster panels, than between them. The assumption is that panels or clusters are independent of one another; that is, there is no correlation between them. The correlation in the data comes from within the panels. This does not always occur though, but because correlation is assumed to exist within panels, standard errors of predictors are assumed to entail correlation, and need to be adjusted using a sandwich variance estimator.

3.4.3 Bootstrapping

Bootstrapping is an entire area of statistics in itself. Here we are discussing bootstrapped standard errors. Statisticians have devised number of ways to bootstrap. I shall develop a function that will bootstrap the model standard errors. I set the number of bootstraps at 100, but it could have been higher for perhaps a bit more accuracy.

```
>library(boot)
> bootmod <- glm(died ~ white + hmo + los + factor(type),
                  family=binomial, data=medpar)
> t <- function ( x, i) {
   xx <- x[i,]
   bsglm <- glm(died ~ white + hmo + los + factor(type),
   family = binomial, data = medpar)
   return(sqrt(diag(vcov(bsglm))))
   }
> bse <- boot(medpar, t, R = 100)
> sqrt(diag(vcov(bootmod)))
> bootse < - apply(bse$t, 2, mean)
```

The bootstrapped standard errors are in the vector, *bootse*. We'll attach them to the table of standard errors to which we keep expanded as we add more types of adjustments.

```
>newcoefse3 <- data.frame( coef, se, scse, rse, bootse)
> round(newcoefse3, 6)
                    coef       se     scse      rse   bootse
(Intercept)    -0.720149 0.219073 0.221302 0.219435 0.219073
white           0.303663 0.209120 0.211248 0.210398 0.209120
hmo             0.027204 0.151242 0.152781 0.150973 0.151242
los            -0.037193 0.007799 0.007878 0.009727 0.007799
factor(type)2   0.417873 0.144318 0.145786 0.145243 0.144318
factor(type)3   0.933819 0.229412 0.231746 0.229306 0.229412
```

The bootstrap algorithm incorporates a great deal of randomness into the calculations, and it generally takes some time to calculate. Each run of the model produces different standard error results. Most analysts appear to prefer employing sandwich standard errors.

3.5 RISK FACTORS, CONFOUNDERS, EFFECT MODIFIERS, AND INTERACTIONS

There are a few terms that are commonly employed when modeling data using logistic regression. Analysts should be aware of their meanings. When a logistic model is being estimated, it is traditionally recognized that the binary variable being modeled is referred to as the *response* term. There is also a single term that is thought to be of foremost interest to the model. This term is known as a *risk factor* and is usually binary or categorical. For instance, suppose a model for which "died within a specific period" is the response, and we have a variable which is a type of physical impairment called *myocardial*; for example, with myocardial = 1 indicating that the subject had an anterior site heart attack, and myocardial = 0 signifying that the damage to the heart is at a nonanterior site. We want to model *died* in order to determine whether anterior or nonanterior site heart attacks have a higher probability of death. We are modeling *died*, but are foremost interested in how levels of *myocardial* bear on the probability of death. *Myocardial* is called a *risk factor.*

The model may also include a predictor which significantly relates to the response, as well as to the risk factor. However, our primary interest is not to learn about this predictor, which is called a *confounder.* The inclusion or exclusion of a confounder has a significant effect on the coefficient of the risk factor.

An *effect modifier* is a predictor that interacts with the risk factor. The risk factor and effect modifier are the main effects terms of an interaction which is used to explain the response. An *interaction* term, of course, is constructed when the levels of one predictor influence the response in a different manner based on the levels of another predictor. Suppose that the response term is *death* and we have predictors *white* and *los*. These are variables in the *medpar* data. If we believe that the probability of death based on length of stay in the hospital varies by racial classification, then we need to incorporate an interaction term of *white* × *los* into the model. The main effects only model is:

```
> summary(y0 <- glm(died~ white+los, family=binomial,
          data=medpar))

Coefficients:
              Estimate Std. Error z value Pr(>|z|)
(Intercept) -0.598683   0.213268  -2.807    0.005    **
white        0.252681   0.206552   1.223    0.221
los         -0.029987   0.007704  -3.893    9.92e-05 ***
```

Note that *los* is significant, but *white* is not. Let's create an interaction of *white* and *los* called *wxl*. We insert it into the model, making sure to include the main effects terms as well.

```
> wxl <- medpar$white * medpar$los

> summary(y1 <- glm(died~ white+los+wxl,
family=binomial, data=medpar))
Coefficients:
             Estimate Std. Error z value Pr(>|z|)
(Intercept) -1.04834    0.28024  -3.741 0.000183 ***
white        0.77092    0.29560   2.608 0.009107 **
los          0.01002    0.01619   0.619 0.535986
wxl         -0.04776    0.01829  -2.611 0.009035 **
```

The interaction term is significant. It makes no difference if the main effects terms are significant or not. Only the interaction term, is interpreted for this model. We calculate the odds ratios of the interaction of *white* and *los* from 1 to 40 as:

$$OR_{interaction} = \exp(\beta_{white} + \beta_{wxl} * los[1:40]) \tag{3.19}$$

That is, we add the slope of the binary predictor to the product of the slope of the interaction and the value(s) of the continuous predictor, exponentiating the whole.

Odds ratios of death for a white patient for length of stay 1–40 days.
Note that odds of death decreases with length of stay.

```
> ior <- exp(0.77092 + (-0.04776*1:40))
> ior
 [1] 2.0609355 1.9648187 1.8731846 1.7858241 1.7025379 1.6231359 1.5474370
 [8] 1.4752685 1.4064658 1.3408718 1.2783370 1.2187186 1.1618807 1.1076936
[15] 1.0560336 1.0067829 0.9598291 0.9150652 0.8723889 0.8317029 0.7929144
[22] 0.7559349 0.7206800 0.6870694 0.6550262 0.6244775 0.5953535 0.5675877
[29] 0.5411169 0.5158806 0.4918212 0.4688839 0.4470164 0.4261687 0.4062933
[36] 0.3873448 0.3692800 0.3520578 0.3356387 0.3199854
```

Interactions for Binary × Binary. Binary × Categorical, Binary × Continuous, Categorical × Categorical, Categorical × Continuous, and Continuous × Continuous may be developed, as well as three-level interactions. See Hilbe (2009) for a thorough analysis of interactions. For now, keep in mind that when incorporating an interaction term into your model, be sure to include the terms making up the interaction in the model, but don't worry about their interpretation or significance. Interpret the interaction based on levels of particular values of the terms. When LOS is 14, we may interpret the odds ratio of the interaction term as:

> *White patients who were in the hospital for 14 days have a some 10% greater odds of death than do non-white patients who were in the hospital for 14 days.*

SAS CODE

```
/* Section 3.1 */

*Refer to the code in section 1.4 to import and print edrelig dataset;
*Refer to proc freq in section 2.4 to generate the frequency table;
*Build logistic model and obtain odds ratio & covariance matrix;
proc genmod data = edrelig descending;
      class educlevel (ref = 'AA') / param = ref;
      model religious = age male kids educlevel/dist = binomial
            link = logit covb;
      estimate "Intercept" Intercept 1 / exp;
      estimate "Age" age 1 / exp;
      estimate "Male" male 1 / exp;
      estimate "Kid" kids 1 / exp;
      estimate "BA" educlevel 1 0 / exp;
      estimate "MA/PhD" educlevel 0 1 / exp;
run;

*Refer to proc iml in section 2.3 and the full code is provided
online;
```

```
/* Section 3.2 */

*Build the logistic model and obtain the deviance residual;
proc genmod data=edrelig descending;
    class educlevel (ref='AA') / param=ref;
    model religious=age male kids educlevel/dist=binomial link=logit;
    output out=residual resdev=deviance;
run;

*Refer to proc means in section 2.5 to summarize deviance residual;
*Build the logistic model and obtain the Person residual;
proc genmod data=edrelig descending;
    class educlevel (ref='AA') / param=ref;
    model religious=age male kids educlevel/dist=binomial link=logit;
    output out=residuals reschi=pearson;
run;

*Pearson Chi2 statistic;
proc sql;
    create table pr as
    select sum(pearson**2) as pchi2, sum(pearson**2)/595 as disp
    from residuals;
quit;

*Refer to proc print in section 2.2 to print dataset pr-Chi2
statistic;
*Build the logistic model and obtain statistic;
proc genmod data=edrelig descending;
    class educlevel (ref='AA') / param=ref;
    model religious=age male kids educlevel/dist=binomial link=logit;
    output out=obstats leverage=hat stdreschi=stdp
    stdresdev=stddev;
run;

*Summary for statistic;
proc means data=obstats min q1 median mean q3 max maxdec=6;
      var hat stdp stddev;
run;
/* Section 3.3 */

*Build the logistic model with class;
proc genmod data=medpar descending;
      class type (ref='1') / param=ref;
      model died=white hmo los type / dist=binomial link=logit covb;
run;

/* Section 3.4 */

*Refer to proc iml in section 2.3 and the full code is provided online;
*Sort the dataset;
proc sort data=medpar;
        by descending type;
run;
```

```
*Use quasilikelihood function to generate scaling standard error;
proc glimmix data=medpar order=data;
    class type;
    model died (event='1')=white hmo los type/dist=binary link=logit
          solution;
    random _RESIDUAL_;
run;

*Generate the robust standard errors;
proc surveylogistic data=medpar;
      class type (ref='1') / param=ref;
      model died (event='1')=white hmo los type;
run;

*Generate the bootstrapped standard errors;
%macro bootstrap (Nsamples);
proc surveyselect data=medpar out=boot
      seed=30459584 method=urs samprate=1 rep=&nsamples.;
run;

proc genmod data=boot descending;
      class type (ref='1') / param=ref;
      model died =white hmo los type / dist=binomial link=logit;
      freq numberhits;
      by replicate;
      ods output ParameterEstimates=est;
run;

data est1;
      set est;
      parameter1=parameter;
      if parameter="Scale" then delete;
      if level1=2 then parameter1="type2";
      else if level1=3 then parameter1="type3";
run;

proc means data=est1 mean;
      class parameter1;
      var StdErr;
run;
%mend;
%bootstrap(100);

/* Section 3.5 */

*Refer to proc genmod in section 1.4 to build the logistic model;
*Build the logistic model with interaction;
proc genmod data=medpar descending;
      model died =white los white*los/ dist=binomial link=logit;
run;

*Generate odds ratios for los from 1 to 40;
data ior;
      do i=1 to 40;
      or=exp(0.7709+(-0.0478*i));
```

```
      output;
      end;
run;

*Refer to proc print in section 2.2 to print dataset ior;
```

STATA CODE

```
3.1
. use edrelig, clear
. glm religious age male kids i.educlevel, fam(bin) nolog nohead eform
. glm religious age male kids i.educlevel, fam(bin) nolog eform

3.2
. e(deviance)                // deviance
. e(deviance_p)              // Pearson Chi2
. e(dispers_p)               // Pearson dispersion
. di e(ll)                   // log-likelihood
. gen loglike=e(ll)
. scalar loglik=e(ll)
. di loglik
. predict h, hat
. sum(h)                     // hat matrix diagonal
. predict stpr, pear stand
. sum stpr                   // stand. Pearson residual
. predict stdr, dev stand
. sum stdr                   // stand deviance residual

3.3
. use medpar, clear
. qui glm died white hmo los i.type, fam(bin)
. estat ic
. abic

3.4
. glm died white hmo los i.type, fam(bin) scale(x2) nolog nohead
. glm died white hmo los i.type, fam(bin) vce(robust) nolog nohead
. glm died white hmo los i.type, fam(bin) vce(boot) nolog nohead

3.5
. glm died white los, fam(bin) nolog nohead
. gen wxl <- white*los
. glm died white los wxl, fam(bin) nolog nohead
. glm died white los wxl, fam(bin) nolog nohead eform
```

Testing and Fitting a Logistic Model

4

4.1 CHECKING LOGISTIC MODEL FIT

4.1.1 Pearson *Chi*2 Goodness-of-Fit Test

I earlier mentioned that the Pearson *Chi*2 statistic, when divided by the residual degrees of freedom, provides a check on the correlation in the data. The idea is to observe if the result is much above the value of 1.0. That is, a well-fitted model should have the values of the Pearson *Chi*2 statistic and residual degrees of freedom closely the same. The closer in value, the better the fit.

$$\frac{\text{Pearson } Chi2}{\text{Residual } dof} \sim 1.0$$

This test, as we shall later discuss, is extremely useful for evaluating extra dispersion in grouped logistic models, but for the observation-based models we are now discussing it is not. A large discrepancy from the value of 1, though, does indicate general extra dispersion or extra correlation in the data, for which use of sandwich or scaled standard errors is an appropriate remedy.

A traditional Pearson *Chi*2 goodness-of-fit (GOF) test, however, is commonly used to assess model fit. It does this by leaving the value of the Pearson *Chi*2 statistic alone, considering it instead to be *Chi*2 distributed with the residual degrees of freedom defining the *Chi2* degrees of freedom. The p-values are based on the distribution, `1-pchisq(pchi2,df)`

*Chi*2(Pearson *Chi*2, *rdof*)

We may code the Pearson *Chi*2 GOF test, creating a little table based on the *mymod* model, as:

```
> pr <- sum(residuals(mymod, type="pearson")^2)
> df <-mymod$df.residual
> p_value <- pchisq(pr, mymod$df.residual, lower=F)
> print(matrix(c("Pearson Chi GOF","Chi2","df","p-value", " ",
+     round(pr,4),df, round(p_value,4)), ncol=2))
     [,1]              [,2]
[1,] "Pearson Chi GOF" " "
[2,] "Chi2"            "1519.4517"
[3,] "df"              "1489"
[4,] "p-value"         "0.2855"
```

This test is still found in many books, articles, and in research reports. Analysts should be aware however, that many statisticians no longer rely on this test as a global fit test. Rather than using a single test to approve or disapprove a model as well fit, statisticians now prefer to employ a variety of tests to evaluate a model. The distributional assumptions upon which tests like this are based are not always met, or are only loosely met, which tends to bias test results. Care needs to be taken when accepting test results.

> With a $p > .05$, the Pearson Chi2 GOF test indicates that we can reject the hypothesis that the model is not well-fitted. In short, we may use the test result to support an acceptance of the model.

4.1.2 Likelihood Ratio Test

In Chapter 2, Section 2.3, we defined the likelihood ratio test as:

$$\text{Likelihood ratio test} = -2\{\mathcal{L}_{\text{reduced}} - \mathcal{L}_{\text{full}}\} \tag{4.1}$$

Using the *drop1* function, an analyst may assess which model of many nested models are better fitted. For example, we create a full model using the *medpar* data. Each predictor is dropped from the model in turn, providing a display of the deviance, Akaike information criterion (AIC), likelihood ratio test statistic, and associated p-value.

```
> summary(mymod <- glm(died ~ white + los + hmo + factor(type),
                        family = binomial,
                        data = medpar))
```

```
Coefficients:
                Estimate    Std. Error z value Pr(>|z|)
(Intercept)     -0.720149   0.219073  -3.287   0.00101 **
white            0.303663   0.209120   1.452   0.14647
los             -0.037193   0.007799  -4.769   1.85e-06 ***
hmo              0.027204   0.151242   0.180   0.85725
factor(type)2    0.417873   0.144318   2.896   0.00379 **
factor(type)3    0.933819   0.229412   4.070   4.69e-05 ***

Null deviance: 1922.9 on 1494 degrees of freedom
Residual deviance: 1881.2 on 1489 degrees of freedom
AIC: 1893.2

> drop1(mymod, test="Chi")
Single term deletions

Model:
died ~ white + los + +hmo + factor(type)
              Df Deviance    AIC     LRT  Pr(>Chi)
<none>            1881.2 1893.2
white          1  1883.3 1893.3  2.1778    0.1400
los            1  1907.9 1917.9 26.7599 2.304e-07 ***
hmo            1  1881.2 1891.2  0.0323    0.8574
factor(type)   2  1902.9 1910.9 21.7717 1.872e-05 ***
```

4.1.3 Residual Analysis

The analysis of residuals plays an important role in assessing logistic model fit. The analyst can see how the model fits rather than simply looking at a statistic. Analysts have devised a number of residuals to view the relationships in a logistic model. Most all residuals that are used in logistic regression were discussed in Chapter 3, Section 3.2, although a couple were not. Table 4.1 summarizes the foremost logistic regression residuals. Table 4.2 gives the R code for producing them. We will only use a few of these residuals in this book, but all have been used in various contexts to analyze the worth of a logistic model. Finally, I shall give the code and graphics for several of the foremost used residual analyses used in publications.

The Anscombe residuals require calculation of an incomplete beta function, which is not part of the default R package. An *ibeta* function is displayed below, together with code to calculate Anscombe residuals for the *mymod* model above. Paste the top lines into the R editor and run. *ans* consists of logistic Anscombe residuals. They are identical to Stata and SAS results.

TABLE 4.1 Residuals for Bernoulli logistic regression

Raw	r	$y-\mu$
Pearson	r^p	$(y-\mu)/\sqrt{\mu(1-\mu)}$
Deviance	r^d	$\sqrt{2\sum\{\ln(1/\mu)\}}$ if $y=1$ $\sqrt{2\sum\{\ln(1/(1-\mu))\}}$ if $y=0$
Stand. Pearson	r^{sp}	$\frac{r^p}{\sqrt{1-h}}$
Stand. deviance	r^{sd}	$\frac{r^d}{\sqrt{1-h}}$
Likelihood	r^l	$\text{sgn}(y-\mu)\sqrt{h(r^p)^2+(1-h)(r^d)^2}$
Anscombe	r^A	$\frac{A(y)-A(\mu)}{\{\mu(1-\mu)\}^{1/6}}$

where $A(z)$ = Beta(2/3, 2/3)*{IncompleteBeta(z, 2/3, 2/3), and $z = (y; \mu)$ Beta(2/3, 2/3) = 2.05339. When $z = 1$, the function reduces to the Beta (see Hilbe, 2009).

Cooks' distance	r^{CD}	$\frac{hr^p}{C_n(1-h)^2}$, C_n = number of coefficients
Delta Pearson	$\Delta Chi2$	$(r^{sd})^2$
Delta deviance	ΔDev	$(r^d)^2+h(r^{sp})^2$
Delta beta	$\Delta\beta$	$\frac{h(r^p)^2}{(1-h)^2}$

```
y <- medpar$died ; mu <- mymod$fitted.value
a <- .666667 ;  b <- .666667
ibeta<- function(x,a,b){ pbeta(x,a,b)*beta(a,b) }
A <- ibeta(y,a,b) ;  B <- ibeta(mu,a,b)
ans <-  (A-B)/  (mu*(1-mu))^(1/6)

> summary(ans)
   Min. 1st Qu.  Median    Mean 3rd Qu.    Max.
-1.5450 -1.0090 -0.9020 -0.0872  1.5310  3.2270
```

Residual analysis for logistic models is usually based on what are known as n-asymptotics. However, some statisticians suggest that residuals should be based on m-asymptotically formatted data. Data in observation-based form; that is, one observation or case per line, are in n-asymptotic format. The datasets we have been using thus far for examples are in n-asymptotic form.

TABLE 4.2 Residual code

```
mu <- mymod$fitted.value                    # predicted probability; fit
r <- medpar$died - mu                       # raw residual
dr <-resid(mymod, type="deviance")          # deviance resid
pr <- resid(mymod, type="pearson")          # Pearson resid
hat <- hatvalues(mymod)                     # hat matrix diagonal
stdr <- dr/sqrt(1-hat)                      # standardized deviance
stpr <- pr/sqrt(1-hat)                      # standardized Pearson
deltadev <- dr^2 + hat*stpr^2               # Δ deviance
deltaChi2 <- stpr^2                         # Δ Pearson
deltaBeta <- (pr^2*hat/(1-hat)^2)           # Δ beta
ncoef <- length(coef(mymod))                # number coefficients
# Cooke's distance
cookD <- (pr^2 * hat) / ((1-hat)^2 * ncoef * summary(mymod)$dispersion)
```

m-asymptotic data occurs when observations with the same values for all predictors are considered as a single observation, but with an extra variable in the data signifying the number of *n*-asymptotic observations having the same covariate pattern. For example, let us consider the 1495 observation *medpar* data for which we have kept only *white*, *hmo,* and *type* as predictors for *died.* There are 1495 observations for the data in this format. In *m*-asymptotic format the data appears as:

```
   white  hmo  type     m
1      0    0     1    72
2      0    0     2    33
3      0    0     3    10
4      0    1     1     8
5      0    1     2     4
6      1    0     1   857
7      1    0     2   201
8      1    0     3    83
9      1    1     1   197
10     1    1     2    27
11     1    1     3     3
```

There are several ways to reduce the three variable subset of the *medpar* data to *m*-asymptotic form. I will show a way that maintains the *died* response variable, which is renamed *dead* due to it not being a binary variable, and then show how to duplicate the above table.

```
> data(medpar)
> test <- subset(medpar, select = c(died, white, hmo, type))
> white <- factor(test$white)
> hmo   <-  factor(test$hmo)
> type  <- factor(test$type)
> mylgg <- na.omit(data.frame(cast(melt(test, measure="died"),
+       white + hmo + type ~ .,
+       function(x) {c(alive=sum(x==0), dead=sum(x==1))} )))
> mylgg$m <- mylgg$alive + mylgg$dead
> mylgg
   white hmo type alive dead   m
1      0   0    1    55   17  72
2      0   0    2    22   11  33
3      0   0    3     6    4  10
4      0   1    1     7    1   8
5      0   1    2     1    3   4
6      1   0    1   580  277 857
7      1   0    2   119   82 201
8      1   0    3    43   40  83
9      1   1    1   128   69 197
10     1   1    2    19    8  27
11     1   1    3     2    1   3
```

The code above produced the 11 covariate pattern *m*-asymptotic data, but I also provide *dead* and *alive*, which can be used for grouped logistic models in the next chapter. *m* is simply the sum of *alive* and *dead*. For example, look at the top line. With $m = 72$, we know that there were 72 times in the reduced *medpar* data for which *white*=0, *hmo*=0, and *type*=1. For that covariate pattern, *died*=1 (dead) occurred 17 times and *died*=0 (alive) occurred 55 times.

To obtain the identical covariate pattern list where only *white*, *hmo*, *type*, and *m* are displayed, the following code reproduces the table.

```
> white <- mylgg$white
> hmo<- mylgg$hmo
> type <- mylgg$type
> m <- mylgg$m
> m_data <- data.frame(white, hmo,type,m)
> m_data
    white  hmo  type    m
1       0    0     1   72
2       0    0     2   33
3       0    0     3   10
4       0    1     1    8
5       0    1     2    4
6       1    0     1  857
```

```
7          1     0      2  201
8          1     0      3   83
9          1     1      1  197
10         1     1      2   27
11         1     1      3    3
```

It should be noted that all but one possible separate covariate pattern exists in this data. Only the covariate pattern, [*white*=0, *hmo*=1, *type*=3] is not part of the *medpar* dataset. It is therefore not in the *m*-asymptotic data format.

I will provide codes for Figures 4.1 through 4.3 that are important when evaluating logistic models as to their fit. Since R's *glm* function does not use a *m*-asymptotic format for residual analysis, I shall discuss the traditional *n*-asymptotic method. Bear in mind that when there are continuous predictors in the model, *m*-asymptotic data tend to reduce to *n*-asymptotic data. Continuous predictors usually have many more values in them than do binary and categorical predictors. A model with two or three continuous predictors typically results in a model where there is no difference between *m*-asymptotic and *n*-asymptotic formats. Residual analysis on observation-based data is the traditional manner of executing the plots, and are the standard way of graphing in R. I am adding *los* (length of stay; number of days in hospital) back into the model to remain consistent with earlier modeling we have done on the *medpar* data.

You may choose to construct residual graphs using *m*-asymptotic methods. The code to do this was provided above. However, we shall keep with the standard methods in this chapter. In the next chapter on grouped logistic models, *m*-asymptotics is built into the model.

R code for creating the standard residuals found in literature related to logistic regression is given in Table 4.2. Code for creating a simple squared standardized deviance residual versus *mu* graphic (Figure 4.1) is given as:

```
data(medpar)
mymod <- glm(died ~ white + hmo + los + factor(type),
             family=binomial, data=medpar)
summary(mymod)
mu <- mymod$fitted.value                   # predicted value;
                                             probability that
                                             died==1
dr <-resid(mymod, type="deviance") # deviance residual
hat <- hatvalues(mymod)                    # hat matrix diagonal
stdr <- dr/sqrt(1-hat)                     # standardized
                                             deviance residual
plot(mu, stdr^2)
abline(h = 4, col="red")
```

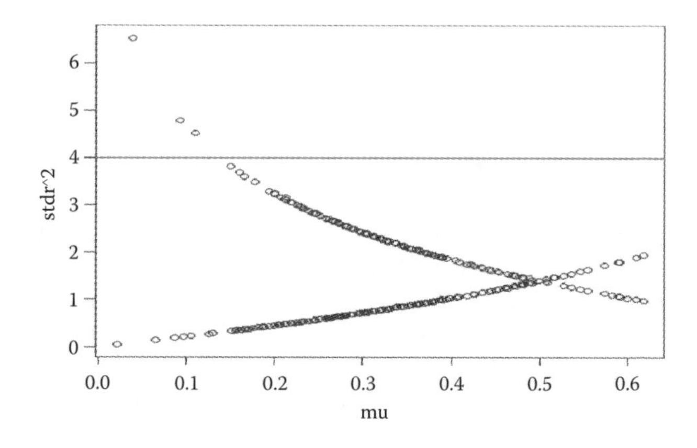

FIGURE 4.1 Squared standardized deviance versus mu.

Analysts commonly use the plot of the square of the standardized deviance residuals versus *mu* to check for outliers in a fitted logistic model. Values in the plot greater than 4 are considered outliers. The values on the vertical axis are in terms of standard deviations of the residual. The horizontal axis are predicted probabilities. All figures here are based on the *medpar* data.

Another good way of identifying outliers based on a residual graph is by use of Anscombe residuals versus *mu*, or the predicted probability that the response is equal to 1. Anscombe residuals adjust the residuals so that they are as normally distributed as possible. This is important when using 2, or 4 when the residual is squared, as a criterion for specifying an observation as an outlier. It is the 95% criterion so commonly used by statisticians for determining statistical significance. Figure 4.2 is not much different from Figure 4.1 when squared standardized deviance residuals are used in the graph. The Anscombe plot is preferred.

```
> plot(mu, ans^2)
> abline(h = 4, lty = "dotted")
```

A leverage or influence plot (Figure 4.3) may be constructed as:

```
> plot(stpr, hat)
> abline(v=0, col="red")
```

Large *hat* values indicate covariate patterns that differ from average covariate patterns. Values on the horizontal extremes are high residuals. Values that are high on the *hat* scale, and low on the residual scale; that is, high in the middle and close to the zero-line do not fit the model well. They are also difficult to detect as influential when using other graphics. There are some seven

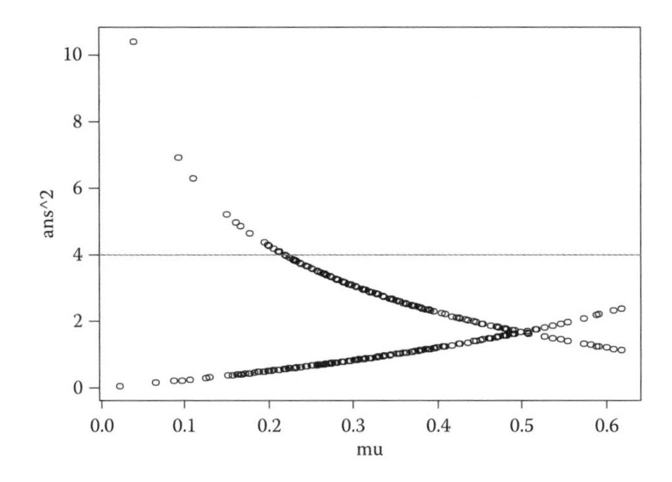

FIGURE 4.2 Anscombe versus mu plot. Values >4 are outliers.

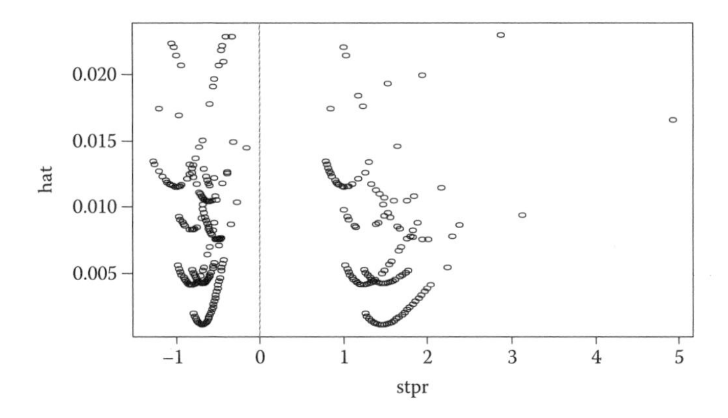

FIGURE 4.3 Leverage plot.

observations that fit this characterization. They can be identified by selecting *hat* values greater than 0.4 and squared residual values of |2|.

A wide variety of graphics may be constructed from the residuals given in Table 4.2. See Hilbe (2009), Bilger and Loughin (2015), Smithson and Merkle (2014), and Collett (2003) for examples.

4.1.4 Conditional Effects Plot

A nice feature of logistic regression is its ability to allow an analyst to plot the predicted probability of an outcome on a continuous predictor, factored across levels of another categorical predictor. Many times the categorical predictor is binary so that two curves of the continuous predictor are displayed, but more than two levels are possible. The conditional effects plot produces this type of graphic, which is quite valuable in understanding the relationship of the probability of the outcome on important predictors.

For an example I use the *medpar* data, modeling the probability of death by length of stay in the hospital, factored by the type of admission. The three levels of the variable *type* in the data are (1) elective, (2) urgent, and (3) emergency. Level 1 has been employed as the reference level. In the conditional effects plot, however, each level of *type* is used to produce a curve of the probability of death for a given length of stay in the hospital. The plot shows three length of stay curves—one for each level of *type*.

The variable *type* is a single variable in the data. The first thing we must do is convert the factor variable into three numeric levels. Then the data are modeled and the predicted values of *los* are created for each level of *type*. I have placed the code in Table 4.3 so that it can be placed in the R [File > "New Script"] editor and executed.

The figure created is displayed as Figure 4.4. The probability curves make sense, with the left most being "elective" admissions, which one expects to have a less lengthy hospital stay than for more serious admission types. "Urgent" admissions is the middle curve and "emergency" the right most, which has both a longer length of stay and also a higher probability of death. When patients are in the hospital for a long time their risk of death is no longer as differentiated.

TABLE 4.3 Code for creating a conditional effects plot

```
data(medpar)
admit <- as.numeric(medpar$type) #convert factor to level values
cep <- glm(died ~ los + admit, data=medpar, family=binomial)
K1 <- coef(cep)[1] + coef(cep)[2]*medpar$los + coef(cep)[3]*1
R1 <- 1/(1+exp(-K1))
K2 <- coef(cep)[1] + coef(cep)[2]*medpar$los + coef(cep)[3]*2
R2 <- 1/(1+exp(-K2))
K3 <- coef(cep)[1] + coef(cep)[2]*medpar$los + coef(cep)[3]*3
R3 <- 1/(1+exp(-K3))

layout(1)
plot(medpar$los, R1, col=1, main='P[Death] within 48 hr admission',
  sub= "Black=1; Red=2; Yellow=3", ylab='Type of Admission',
  xlab='LOS', ylim=c(0,0.4))
lines(medpar$los, R2, col=2, type='p')
lines(medpar$los, R3, col=3, type='p')
```

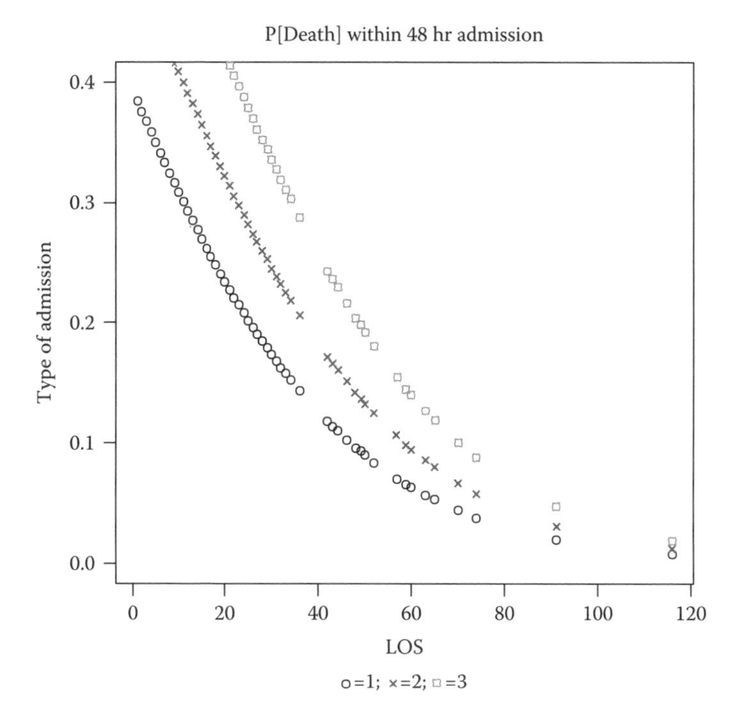

FIGURE 4.4 Conditional effects plot.

4.2 CLASSIFICATION STATISTICS

Logistic regression is many times used as a classification tool. However, it should be noted that the ability to classify or discriminate between the two levels of the response variable is due more to the degree of separation between the levels and size of the regression coefficients than it is to the logistic model itself. Discriminate analysis and other classification schemes can also do a good job in classifying and are not logistic models. On the other hand, logistic models are easy to work with and are robust in the classification results they provide the analyst.

There are three basic or standard types of classification tools used with logistic regression—the sensitivity–specificity (S–S) plot, the receiver operator characteristic (ROC) curve, and a confusion table. We will address each of these in this section. Each of these tests is based on a cut point, which determines the optimal probability value with which to separate predicted versus observed successes (1) or failures (0).

For an example we shall continue with the model used for residual analysis earlier in the chapter. It is based on the *medpar* data, with *died* as the response variable and *white*, *hmo*, *los* and levels of *type* as the predictors. We then obtain predicted probabilities that *died* = =1, which is the definition of *mu*. The goal is then to determine how well the predicted probabilities actually predict classification as *died* = =1, and how well they predict *died* = =0. Analysts are not only interested in correct prediction though, but also in such issues as what percentage of times does the predictor incorrectly classify the outcome. I advise the reader to remember though that logistic models that classify well are not always well-fitted models. If your interest is strictly to produce the best classification scheme, do not be as much concerned about model fit. In keeping with this same logic, a well-fitted logistic model may not clearly differentiate the two levels of the response. It's valuable if a model accomplishes both fit and classification power, but it need not be the case.

Now to our example model:

```
> mymod <- glm(died ~ white + hmo + los + factor(type),
                family=binomial,
                data=medpar)

> mu <- predict(mymod, type="response")
> mean(medpar$died)
[1] 0.3431438
```

Analysts traditionally use the mean of the predicted value as the cut point. Values greater than 0.3431438 should predict that *died* = =1; values lower should predict *died* = =0. For confusion matrices, the mean of the response, or mean of the prediction, will be a better cut point than the default 0.5 value set with most software. If the response variable being modeled has substantially more or less 1's than 0's, a 0.5 cut point will produce terrible results. I shall provide a better criterion for the cut point shortly, but the mean is a good default criterion.

Analysts can use the percentage at which levels of *died* relate to *mu* being greater or less than 0.3431438 to calculate such statistics as specificity and sensitivity. These are terms that originate in epidemiology, although tests like the ROC statistic and curve were first derived in signal theory. Using our example, we have patients who died (D) and those who did not (~D). The probability of being predicted to die given that the patient has died is called model *sensitivity*. The probability of being predicted to stay alive, given the fact that the patient remained alive is referred to as model *specificity*. In epidemiology, the term sensitivity refers to the probability of testing positive for having a disease given that the patient in fact has the disease. Specificity refers to when a patient tests negative for a disease when they in fact do not have the disease. Terms such as *false positive* refers to when a patient tests positive for a disease even though they do not have it. *False negatives* happen when a patient tests negative for a disease, even though they actually have it. These are all important statistics in classification analysis, but model sensitivity and specificity are generally regarded as the most important results. However, false positive and false negative are used with the main statistics for creating the ROC curve. Each of these statistics can easily be calculated from a confusion matrix. All three of these classification tools intimately relate with one another.

The key point is that determining the correct cut point provides the grounds for correctly predicting the above statistics, given an estimated model. The cut point is usually close to the mean of the predicted values, but is not usually the same value as the mean. Another way of determining the proper cut point is to choose a point at which the specificity and sensitivity are closest in values. As you will see though, formulae have been designed to find the optimal cut point, which is usually at or near the site where the sensitivity and specificity are the closest.

All three classification tests are part of the *ROC_test* and *confusion_stat* functions that come with this text. They are located on the book's web site and in the LOGIT package on CRAN. Load both functions since they are required for all three classification tests. We load the data and model as well.

```
> source("C://Rfiles/ROC_test.R")        # or use if already
                                           in memory
> source("C://Rfiles/confusion_stat.R") # or use if already
                                           in memory
> library(COUNT)
> data(medpar)
> mymod <- glm(died ~ los + white + hmo + factor(type),
              family=binomial, data=medpar)
```

We shall start with the S–S plot, which is typically used to establish the cut point used in ROC and confusion matrix tests. The cut point used in *ROC_test* is based on Youden's J statistic (Youden, 1950). The optimal cut point is defined as the threshold that maximizes the distance to the identity (diagnonal) line of the ROC curve. The optimality criterion is based on:

```
max(sensitivities + specificities)
```

Other criteria have been suggested in the literature. Perhaps the most noted alternative is:

```
min((1 - sensitivities)^2 + (1- specificities)^2)
```

Both criteria give remarkably close cut points.

4.2.1 S–S Plot

An S–S plot is a graph of the full range of sensitivity and specificity values that occur for cut-point values ranging from 0 to 1. The intersection of sensitivity and specificity values indicates the point at which the two statistics are closest in value. The graphic, indicates that 0.364 is that point. The associated $cut statistic that is displayed in the non-graphical output is more exact, having a value of 0.363812. This is the cut point we shall later use with the confusion matrix, from which false-positive and related statistics may be calculated (Figure 4.5).

```
> out1< -ROCtest(modeltest,10,type="Sensitivity")
> out1

$cut
[1] 0.363812
```

4.2.2 ROC Analysis

Receiver operator characteristic curves are generally used when statisticians wish to determine the predictive power of the model. It is also used for classification purposes. The ROC curve is understood as the optimal relationship of the model sensitivity by one minus the specificity.

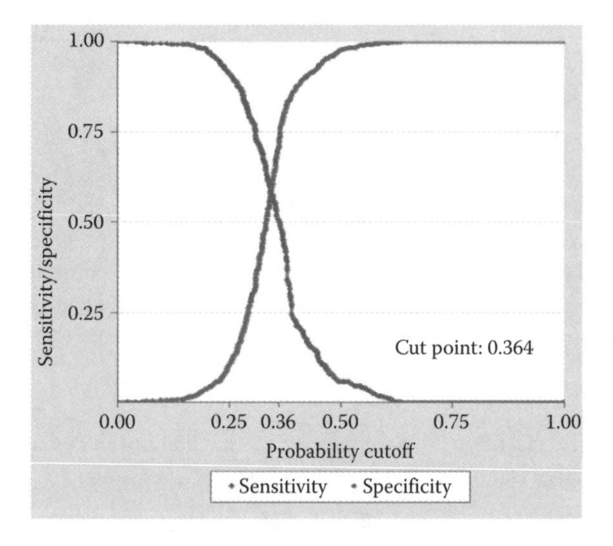

FIGURE 4.5 Sensitivity–specificity plot.

When using ROC analysis, the analyst should look at both the ROC statistic as well as at the plot of the sensitivity versus one minus the specificity. A model with no predictive power has a slope of 1. This represents an ROC statistic of 0.5. Values from 0.5 to 0.65 have little predictive power. Values from 0.65 to 0.80 have moderate predictive value. Many logistic models fit into this range. Values greater than 0.8 and less than 0.9 are generally regarded as having strong predictive power. Values of 0.9 and greater indicate the highest amount degree of predictive power, but models rarely achieve values in this range. The model is a better classifier with greater the values of the ROC statistic, or area under the curve (AUC). Be aware of over-fitting with such models. Validating the model with a validation sample or samples is recommended. See Hilbe (2009) for details.

ROC is a test on the response term and fitted probability. SAS users should note that the ROC statistic described here is referred to Harrell's C statistic.

The *ROCtest* function is used to determine that the predictive power of the model is 0.607. Note that the *type* = *"ROC"* option is given to obtain the test statistic and graphic. Due to the sampling nature of the statistics, the cut point for the ROC curve differs slightly from that of the S–S plot (Figure 4.6).

```
> out2 <- ROCtest(mymod, fold=10, type="ROC")
> out2

$cut
[1] 0.3614538
```

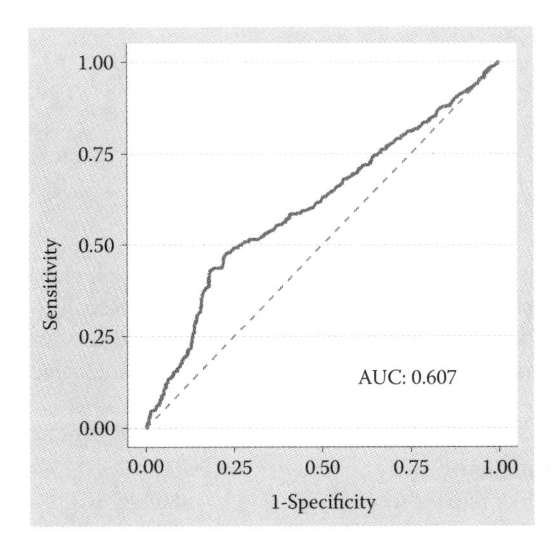

FIGURE 4.6 Receiver operator characteristic curve.

Note that a cutoff value of 0.3615 is used for the AUC statistic. Given that *died* indicates that a patient died within 48 h of admission to the hospital, the AUC statistic can be interpreted as follows: The estimated probability is 0.61 that patients who die have a higher probability of death (higher *mu*) than patients who are alive. This value is very low. A ROC statistic of 0.5 indicates that the model has no predictive power at all. For our model there is some predictive power, but not a lot.

4.2.3 Confusion Matrix

The traditional logistic regression classification table is given by the so-called *confusion matrix* of correctly and incorrectly predicted fitted values. The matrix may be obtained following the use of the previous options of *ROC_test* by typing

```
> confusion_stat(out1$Predicted,out1$Observed)
```

A confusion matrix of values is immediately displayed on screen, together with values for correctly predicted (accuracy), sensitivity, and specificity. The cut point from the S–S plot is used as the confusion matrix cut point.

```
$matrix
     obs    0    1  Sum
pred
0         794  293 1087
1         188  220  408
Sum       982  513 1495

$statistics
   Accuracy Sensitivity Specificity
  0.6782609   0.4288499   0.8085540
```

Other statistics that can be drawn from the confusion matrix and that can be of value in classification analysis are listed below. Recall from earlier discussion that D = patient died within 48 h following admission (outcome = 1) and ~D = patient not died within 48 h after admission (outcome = 0).

Positive predictive value	:	220/408=0.5392157 = 53.92%
Negative predictive value	:	794/1087 = 0.7304508 = 73.05%
False-positive rate for true ~D	:	188/982 = 0.1914460 = 19.14%
False-positive rate for true D	:	293/513 = 0.5711501 = 57.12%
False-positive rate for classified positives	:	188/408 = 0.4607843 = 46.08%

False-negative rate for classified negatives : 293/1087 = 0.2695482 = 26.95%

An alternative way in which confusion matrices have been constructed is based on the closeness in sensitivity and specificity values. That is, either the analyst or an algorithm determines when the sensitivity and specificity values are closest, and then constructs a matrix based on the implications of those values. An example of this method can be made from the *PresenceAbsence* package and function. The cut point, or threshold, is 0.351, which is not much different from the cut point of 0.3638 we used in the *ROC_test* function. The difference in matrix values and the associated sensitivity and specificity values are rather marked though. I added the marginals to provide an easier understanding of various ancillary statistics which may be generated from the confusion matrix.

```
> library(PresenceAbsence)
> mymod <- glm(died ~ white + hmo + los + factor(type),
+            family=binomial, data=medpar)
> mu <- predict(mymod, type="response")
> cmxdf <- data.frame(id=1:nrow(medpar), died=as.
  vector(medpar$died),
+    pred=as.vector(mu))
> cmx(cmxdf, threshold=0.351,which.model=1)
# a function in PresenceAbsence

         Observed    Total
predicted   1   0
          1 292 378    670
          0 221 604    825
Total       513 982   1495
```

The correctly predicted value, or accuracy, is (292 + 604)/1495 or 59.93%. Sensitivity is 292/(292 + 221) or 56.72% and specificity is 604/(378 + 604) or 61.51%. Note that the sensitivity (56.72%) and specificity (61.51%) are fairly close in values—they are as close as we can obtain. If we use the same algorithm with the cut point of 0.363812 calculated by the S–S plot using the criterion described at the beginning of this section, the values are

```
> cmx(cmxdf, threshold=0.363812,which.model=1)

          observed
predicted   1   0    Total
          1 252 233     485
          0 261 749    1010
Total       513 982    1495
```

with an accuracy of 66.96%, a sensitivity of (252/513) 49.12%, and a specificity of (749/982) 76.27%. The small difference in cut point results in a sizeable difference in classification values. The values of accuracy, sensitivity, and specificity obtained using the S–S plot criterion are similar to the values obtained in our last matrix with the same cut point: accuracy = 67.83%, sensitivity = 42.88%, and specificity = 80.86%. Given the variability in results due to sampling, these results can be said to be the same.

Classification is a foremost goal of logistic modeling for those in industries such as credit scoring, banking, ecology, and even astronomy, to name a few. I refer you to Hilbe (2009), Bilger and Loughin (2015), or De Souza et al. (2015) for additional details regarding these tests.

4.3 HOSMER–LEMESHOW STATISTIC

The Hosmer–Lemeshow (H–L) test was designed by their creators as a GOF test to assess the differences between the observed and expected or predicted probabilities as categorized across levels of predicted values. The predicted probabilities of the model are divided into a specified number of groups—usually 10. That is, the range of predicted probabilities is collapsed into 10 groups or quantiles of probabilities. Each group is a range of probabilities. The observed number of 0s and 1s are calculated for each group, and are compared to the count of predicted probabilities of 0s and 1s for each group. The absolute differences are summed, resulting in an H–L *Chi*2 statistic. The degrees of freedom are the number of groups less two. The *p*-values greater than 0.05 indicate a well-fitted model. One rejects the null hypothesis that the observed and predicted probabilities are the same.

The H–L *Chi*2 test is very sensitive to the way in which tied values are handled. Various softer implementations handle ties in different ways. Due to this the H–L statistic and *p*-values for two versions of the test may differ, sometimes by quite a bit. It is also important to have at least five observations in each group in order to generate a meaningful *Chi*2 statistic. When this is not the case, reduce the number of groups to 8, or perhaps 6. I suggest that you use the statistic three times, with 8, 10, and 12 groups for moderate to large sized data sets. Check to determine if the tests show a well-fitted, or not well-fitted, model for all three groups.

The *medpar* data will be used for an example of this test. I first display the summary results of a logistic model:

```
> data(medpar)
> summary(glm(died ~ white + los  + hmo + factor(type),
          family = binomial, data = medpar))
```

```
Coefficients:
                 Estimate Std. Error z value Pr(>|z|)
(Intercept)     -0.720149   0.219073  -3.287  0.00101 **
white            0.303663   0.209120   1.452  0.14647
los             -0.037193   0.007799  -4.769 1.85e-06 ***
hmo              0.027204   0.151242   0.180  0.85725
factor(type)2    0.417873   0.144318   2.896  0.00379 **
factor(type)3    0.933819   0.229412   4.070 4.69e-05 ***
---
    Null deviance: 1922.9  on 1494  degrees of freedom
Residual deviance: 1881.2  on 1489  degrees of freedom
AIC: 1893.2
```

We next load the source code for the H–L *Chi*2 test into memory. I show that the code is stored in the *c://Rfiles* directory, but you should change this to where you have placed the function. The *HLTest.R* file is on the book's web site. It is from Bilger and Loughin (2015).

```
> source("c://Rfiles/HLTest.R")       # on book's web site
> HLChi10 <- HLTest(obj = mymod,g= 10)
> cbind(HLChi10$observed, round(HLChi10$expect, digits = 1))
                  Y0  Y1 Y0hat Y1hat
[0.0219,0.252]   108  43 119.1  31.9
(0.252,0.289]    106  42 107.7  40.3
(0.289,0.31]     116  40 109.3  46.7
(0.31,0.328]     106  37  97.7  45.3
(0.328,0.343]    123  43 110.5  55.5
(0.343,0.354]    113  38  98.1  52.9
(0.354,0.371]    124  41 104.8  60.2
(0.371,0.388]     53 105  97.3  60.7
(0.388,0.445]     53  47  58.9  41.1
(0.445,0.618]     80  77  78.5  78.5
> HLChi2

    Hosmer and Lemeshow goodness-of-fit test with 10 bins

data:  mymod
X2 = 82.8324, df = 8, p-value = 1.31e-14
```

The *p*-value approximates 0, which is far under the criterion of 0.05. It appears that the model is not well fitted. We can inspect the relationships between Y1 and Y1Hat (observed *died* and associated predicted *died*, or *mu*), and between Y0 and Y0hat (observed *died* = 0, and corresponding predicted *died*). Some of the pairs are close, but many are not. One of the worst fitted pairs are in the probability range 0.371–0.388, with Y1–Y1hat as 105–60.7, and Y0–Y0hat at 53–97.3. The second group, 0.252–0.289, is well fitted: Y1–Y1hat 42–40.3 and Y0–Y0hat 106–107.7. But too many groups are marginal to poor.

If we divide up the response probability space into 12 divisions the results appear as:

```
> HLChi12 <- HLTest(obj = mymod,g= 12)
> HLChi12

    Hosmer and Lemeshow goodness-of-fit test with 12 bins

data:  mymod
X2 = 84.8001, df = 10, p-value = 5.718e-14

> cbind(HL$observed, round(HL$expect, digits = 1))
                  Y0 Y1 Y0hat Y1hat
[0.0219,0.246]    87 38  99.6  25.4
(0.246,0.278]     94 31  92.0  33.0
(0.278,0.297]     97 37  95.1  38.9
(0.297,0.313]    103 38  97.5  43.5
(0.313,0.329]    101 35  91.8  44.2
(0.329,0.343]     77 26  68.2  34.8
(0.343,0.354]    113 38  98.1  52.9
(0.354,0.362]     75 15  57.5  32.5
(0.362,0.38]      79 55  84.0  50.0
(0.38,0.391]      35 80  70.4  44.6
(0.391,0.454]     62 58  69.0  51.0
(0.454,0.618]     59 62  58.7  62.3
```

The *Chi*2 test again indicates that the model is ill fitted.

In order to show how different code can result in different results, I used code for the H–L test in Hilbe (2009). Rather than groups defined and displayed by range, they are calculated as ranges, but the mean of the groups is displayed in output. The number of observations in each group is also given.

This code will develop three H–L tables, with 8, 10, and 12 groups. The 12 group table is displayed below.

```
> source('HL_GOFtest.r')    # on book's web site
> medpar2<- na.omit(medpar) # drop obs with missing value(s)
> hlGOF.test(medpar2$died, predict(mymod, medpar2,
            type='response'), breaks=12)

For # Cuts = 12    # Data = 1495
Cut # Total #Patterns # Resp. # Pred. Mean Resp. Mean Pred.
1       125      61        38      25.39    0.30400    0.20311
2       124      24        31      32.72    0.25000    0.26384
3       125      14        35      36.16    0.28000    0.28929
```

```
4       124     15      34      38.05   0.27419   0.30689
5       125     11      31      40.10   0.24800   0.32079
6       125      9      33      41.76   0.26400   0.33409
7       124      7      29      43.16   0.23387   0.34806
8       125      5      26      44.80   0.20800   0.35843
9       124     11      44      46.08   0.35484   0.37160
10      125     10      89      48.33   0.71200   0.38660
11      124     20      59      52.32   0.47581   0.42191
12      125     32      64      64.14   0.51200   0.51310
Total # Data: 1495  Total over cuts: 1495
Chisq: 91.32444   d.f.: 10  p-value:  0.00000
```

The *p*-value again tells us that the model is not well fitted. The statistics are similar, but not identical to the table shown earlier. The H–L test is nice summary test to use on a logistic model, but interpret it with care.

4.4 MODELS WITH UNBALANCED DATA AND PERFECT PREDICTION

When the data set you wish to model has few observations, few predictors, and are categorical in nature, it is possible that perfect prediction exists between the predictors and response. That is, for a given covariate pattern only one outcome occurs. Maximum likelihood estimation does not work well in such circumstances. One or more of the coefficients become very large, and standard errors may explode to huge sizes as well. Coefficient values may also be displayed with no value given. When this occurs it is nearly always the case that perfect prediction exists in the data.

Consider a real data set consisting of HIV drug data. The response is given as the number of patients in a study who became infected with HIV. There are two predictors, *cd*4 and *cd*8, each with three levels–0, 1, and 2. The data is weighted by the number of cases having the same pattern of covariates; that is, with the values of *cd*4 and *cd*8 the same.

The data, called *hiv*, is downloaded into R's memory from its original format as a Stata data set.

```
> library(Hmisc)
> hiv <- stata.get("C://ado/hivlgold.dta")
> hiv
   infec cases cd4 cd8
1      0     3   0   0
2      0     8   1   1
```

```
3        0     2   2   2
4        0     5   1   0
5        0     2   2   0
6        0    13   2   1
7        1     1   0   2
8        1     2   1   2
9        1     4   0   0
10       1     4   1   1
11       1     1   2   2
12       1     2   1   0
```

Next, we model the data as a weighted logistic regression. Level 0 of both *cd*4 and *cd*8 are the reference levels.

```
> myhiv <- glm(infec ~ factor(cd4) + factor(cd8),
+                family=binomial, weights=cases, data=hiv)
> summary(myhiv)

                      .    .    .
Coefficients:
              Estimate Std. Error z value Pr(>|z|)
(Intercept)     0.2877     0.7638   0.377    0.706
factor(cd4)1   -1.2040     1.1328  -1.063    0.288
factor(cd4)2  -20.3297  2501.3306  -0.008    0.994
factor(cd8)1    0.2231     1.0368   0.215    0.830
factor(cd8)2   19.3488  2501.3306   0.008    0.994

    Null deviance: 57.251  on 11  degrees of freedom
Residual deviance: 37.032  on  7  degrees of freedom
AIC: 47.032
```

Look at the highest level of both *cd*4 and *cd*8. The coefficient values are extremely high compared to level 2, and the standard errors of both are over 100 times greater than their associated coefficient. None of the Wald *p*-values are significant. The model appears to be ill fitted, to say the least.

Penalized logistic regression was developed to resolve the problem of perfect prediction. Heinze and Schemper (2002) amended a method designed by David Firth (1993) to solve the so-called "problem of separation," which results in at least one parameter becoming infinite, or very large compared to other predictors or levels of predictors in a model. See Hilbe (2009) for a discussion of the technical details of the method.

The same data as above are modeled using Firth's penalized logistic regression. The function, *logistf*() is found in the *logistf* package on CRAN.

```
> firth <- logistf(infec ~ factor(cd4) + factor(cd8), weights=cases, data=hiv)
> firth
```

```
logistf(formula = infec ~ factor(cd4) + factor(cd8), data = hiv,
    weights = cases)
Model fitted by Penalized ML
Confidence intervals and p-values by Profile Likelihood

                    coef  se(coef) lower 0.95 upper 0.95       Chisq           p
(Intercept)    0.2431531 0.7556851 -1.1725243  1.7422757 0.11676459 0.732570400
factor(cd4)1  -1.0206696 1.0903625 -3.2183884  0.9721743 0.98925098 0.319925511
factor(cd4)2  -4.0139131 1.7546659 -9.1437134 -1.1309059 8.42543351 0.003700084
factor(cd8)1   0.1320063 0.9852859 -1.7257106  2.1418014 0.01905676 0.890203896
factor(cd8)2   3.2265668 1.7200153  0.4644354  8.3696176 5.57992109 0.018167541

Likelihood ratio test=16.42534 on 4 df, p=0.00249844, n=47
```

The coefficients appear normal with nothing out of the ordinary. Interestingly the p-values of the second level of *cd*4 and *cd*8, which failed in standard logistic regression, are statistically significant for the penalized logit model. The likelihood ratio test informs us that the penalized model is also not well fitted.

Penalized logistic regression many times produces significant results when standard logistic regression does not. If you find that there is perfect prediction in your model, or that the data is highly unbalanced; for example, nearly all 1s or 0s for a binary variable, penalized logistic regression may be the only viable way of modeling it. Those analysts who model mostly small data sets are more likely to have separation problems than those who model larger data.

4.5 EXACT LOGISTIC REGRESSION

Exact logistic regression is a method of constructing the Bernoulli distribution such that it is completely determined. The method is unlike maximum likelihood or iteratively reweighted least squares (IRLS) which are asymptotic methods of estimation. The model coefficients and p-values are accurate, but at a cost of involving a large number of permutations.

Exact logistic and exact Poisson regressions were originally written for the Cytel Corporation product named LogXact in the late 1990s. SAS, SPSS, and Stata statistical software soon incorporated the procedures into their commercial packages. R's *elrm* package is the closest R has come to providing R users with this functionality. It is a difficult function to employ and I have not been able to obtain results similar to those of the other packages. I will use Stata's *exlogistic* command for an example of the method and its results. The SAS version of the code is at the end of this chapter; the results are the same as Stata output.

Exact logistic regression is typically used by analysts when the size of the data being modeled is too small to yield well-fitted results. It is also used when the data are ill balanced, however, it is not to be used when there is perfect prediction in the model. When that occurs penalized logistic regression should be used—as we discussed in the previous section.

For an example of exact logistic regression, I shall use Arizona hospital data collected in 1991. The data consist of a random sample of heart procedures referred to as CABG and PTCA. CABG is an acronym meaning coronary artery bypass grafting surgery and PTCA refers to percutaneous transluminal coronary angioplasty. It is a nonsurgical method of placing a type of balloon into a coronary artery in order to clear blockage caused by cholesterol. It is a substantially less severe procedure than CABG. We will model the probability of death within 48 h of the procedure on 34 patients who sustained either a CABG or PTCA. The variable *procedure* is 1 for CABG and 0 for PTCA. It is adjusted in the model by the type of admission. *Type* = 1 is an emergency or urgent admit, and 0 is an elective admission. Other variables in the data are not used. Patients are all older than 65.

```
> azheart <- stata.get("c://ado/azcabgptca34.dta")
> head(azheart)
     died procedure age gender los     type
1    Died      CABG  65   Male  10 Elective
2 Survive      CABG  69   Male   7 Emer/Urg
3 Survive      PTCA  76 Female   7 Emer/Urg
4 Survive      CABG  65   Male   8 Elective
5 Survive      PTCA  69   Male   1 Elective
6 Survive      CABG  67   Male   7 Emer/Urg
```

A cross-tabulation of *died* on *procedure* is given as:

```
> library(Hmisc)
> table(azheart$died, azheart$procedure)

          PTCA CABG
  Survive   19    9
  Died       1    5
```

It is clear from the tabulation that more patients died having a CAGB than with a PTCA. A table of *died* on *type* of admission is displayed as:

```
> table(azheart$died, azheart$type)

          Elective Emer/Urg
  Survive       17       11
  Died           4        2
```

First we shall use a logistic regression to model *died* on *procedure* and *type*. The model results are displayed in terms of odds ratios and associated statistics.

```
> exlr <- glm(died ~ procedure + type, family=binomial,
              data=azheart)
> source("c://rfiles/toOR.R")
> toOR(exlr)
                     or   delta  zscore pvalue exp.loci. exp.upci.
(Intercept)      0.0389  0.0482 -2.6170 0.0089    0.0034    0.4424
procedureCABG   12.6548 15.7958  2.0334 0.0420    1.0959  146.1267
typeEmer/Urg     1.7186  1.9296  0.4823 0.6296    0.1903   15.5201
```

Note that there appears to be a statistically significant relationship between the probability of death and type of procedure ($p = 0.0420$). Type of admission does not contribute to the model. Given the size of the data and adjusting for the possibility of correlation in the data we next model the same data as a "quasibinomial" model. Earlier in the book I indicated that the *quasibinomial* option is nothing more than scaling (multiplying) the logistic model standard errors by the square root of the Pearson dispersion statistic.

```
> exlr1 <- glm(died ~ procedure + type, family=quasibinomial,
               data=azheart)
> toOR(exlr1)
                     or   delta  zscore pvalue exp.loci. exp.upci.
(Intercept)      0.0389  0.0478 -2.6420 0.0082    0.0035    0.4324
procedureCABG   12.6548 15.6466  2.0528 0.0401    1.1216  142.7874
typeEmer/Urg     1.7186  1.9114  0.4869 0.6264    0.1943   15.2007
```

The *p*-value of *procedure* is further reduced by scaling. The same is the case when the standard errors are adjusted by sandwich or robust variance estimators (not shown). We might accept *procedure* as a significant predictor of the probability of death—if it were not for the small sample size. If we took another sample from the population of procedures would we have similar results? It is wise to set aside a validation sample to test our primary model. But suppose that we do not have access to additional data? We subject the data to modeling with an exact logistic regression. The Stata code and output are given below.

```
. exlogistic died procedure type, nolog

Exact logistic regression            Number of obs =        34
                                     Model score   =  5.355253
                                     Pr >= score   =    0.0864
----------------------------------------------------------------
died       Odds Ratio Suff. 2*Pr(Suff.)  [95% Conf. Interval]
----------------------------------------------------------------
procedure  10.33644     5     0.0679      .8880104    589.8112
type        1.656699    2     1.0000      .1005901    28.38678
----------------------------------------------------------------
```

The results show that *procedure* is not a significant predictor of *died* at the $p = 0.05$ criterion. This should not be surprising. Note that the odds ratio of procedure diminished from 12.65 to 10.33. In addition, the model *score* statistic given in the header statistics informs us that the model does not fit the data well ($p > 0.05$).

When modeling small and/or unbalanced data, it is suggested to employ exact statistical methods if they are available.

4.6 MODELING TABLE DATA

We have touched on modeling data in table format at various points in the book. The subject is not discussed in other texts on logistic regression, but the problem comes up in the real life experience of many analysts. I shall therefore discuss data that are recorded in tables and how one should best convert it to a format suitable for modeling as a logistic model. We will start with the simplest table, a two-by-two table.

Suppose we have table data in the generic form below:

Table Format

		x	
		0	1
y	0	4	5
	1	6	8

This table has two variables, y and x. It is in summary form. That is, the above table is a summary of data and can be made into two variables when put into the following format.

Grouped Format

y	x	count
0	0	4
0	1	5
1	0	6
1	1	8

The cell ($x = 0$; $y = 0$), or (0,0) in the above table has a value of 4; the cell ($x = 1$; $y = 1$), or (1,1) has a value of 8. This indicates that if the data were in observation-level form, there would be four observations having a pattern of

x,y values of 0,0. If we are modeling the data, with y as the binary response and x as a binary predictor, the observation-level data appears as:

Observation-Level Format

	y	x
1.	0	0
2.	0	0
3.	0	0
4.	0	0
5.	0	1
6.	0	1
7.	0	1
8.	0	1
9.	0	1
10.	1	0
11.	1	0
12.	1	0
13.	1	0
14.	1	0
15.	1	0
16.	1	1
17.	1	1
18.	1	1
19.	1	1
20.	1	1
21.	1	1
22.	1	1
23.	1	1

The above data give us the identical information as we have in the "y-x count" table above it, as well as in the initial table. Each of these three formats yield the identical information. If the analyst simply sums the values of the numbers in the cells, or sums the values of the *count* variable, he/she will know the number of observations in the observation-level data set. $4 + 5 + 6 + 8$ indeed sums to 23.

Note that many times we see table data converted to grouped data in the following format:

y	x	count
1	1	8
1	0	6
0	1	5
0	0	4

I tend to structure grouped data in this manner. But as long as an analyst is consistent, there is no difference in the methods. What is important to remember is that if there are only two binary variables in a table, y and x, and if y is the response variable to be modeled, then it is placed as the left-most column with p^2 levels. p is the number of binary variables, in this case $2^2 = 4$.

The data in grouped format are modeled as a frequency weighted regression. Since y is binary, it will be modeled as a logistic regression, although it also may be modeled as a probit, complimentary loglog, or loglog regression. The key is to enter the counts as a frequency weight.

```
> y <- c(1,1,0,0)
> x <- c(1,0,1,0)
> count <- c(8,6,5,4)
> mydata <- data.frame(y,x,count)

> mymodel <- glm(y ~ x, weights=count, family=binomial,
                 data=mydata)
> summary(mymodel)

                    . . .
Coefficients:
            Estimate Std. Error z value Pr(>|z|)
(Intercept)  0.40547    0.64550   0.628     0.53
x            0.06454    0.86120   0.075     0.94

    Null deviance: 30.789  on 3  degrees of freedom
Residual deviance: 30.783  on 2  degrees of freedom
AIC: 34.783
```

The logistic coefficients are $x = 0.06454$ and intercept as 0.40547. Exponentiation gives the following values:

```
> exp(coef(mymodel))
(Intercept)           x
   1.500000    1.066667
```

To check the above calculations the odds ratio may be calculated directly from the original table data as well. Recall that the odds ratio of predictor x is the ratio of the odds of $y = 1$ divided by the odds of $y = 0$. The odds of $y = 1$ is the ratio of $x = 1$ to $x = 0$ when $y = 1$, and the odds of $y = 0$ is the ratio of $x = 1$ to $x = 0$ when $y = 0$.

		x	
		0	1
y	0	4	5
	1	6	8

```
> (8/5)/(6/4)
[1] 1.066667
```

which is the value calculated as *x* above. Recalling our discussion earlier in the text, the intercept odds is the denominator of the ratio we just calculated to determine the odds ratio of *x*.

```
> 6/4
[1] 1.5
```

which confirms the calculation from R.

When tables are more complex the same logic used in creating the 2×2 table remains. For instance, consider a table of summary data that relates the pass–failure rate among males and females in an introductory to statistics course at Noclue University. The goal is to determine if studying for the final or going to a party or just sleeping instead had a bearing on passing. There are 18 males and 18 females, for a class of 36.

	Gender					
	Female			Male		
	sleep	party	study	sleep	party	study
Grade fail	3	4	2	2	4	3
Grade pass	2	1	6	3	2	4

The data have a binary response, *Grade*, with levels of Fail and Pass, *Gender* has two levels (Female and Male) and student *Type* has three levels (sleep, party, and study). I suggest that the response of interest, *Pass,* be giving the value of 1, with *Fail* assigned 0. For *Gender*, Female = 0 and Male = 1. *Type*: Sleep = 1, Party = 2, and Study = 3. Multiply the levels for the total number of levels or groups in the data. 2 * 2 * 3 = 12. The response variable then will have six 0s and six 1s. When a table has predictors with more than two levels, I recommend using the 0,1 format for setting up the data for analysis.

A binary variable will split its values between the next higher level. Therefore, *Gender* will have alternating 0s and 1s for each half of *Grade.* Since *Type* has three levels, 1–2–3 is assigned for each level of *Gender*. Finally, assign the appropriate count value to each pattern of variables. The first level represents Grade = Fail; Gender = Female; Type = Sleep. We move from the upper left of the top row across the columns of the row, then move to the next row.

```
      Grade   Gender    Type    Count
1:      0       0        1        3
2:      0       0        2        4
3:      0       0        3        2
4:      0       1        1        2
5:      0       1        2        4
6:      0       1        3        3
7:      1       0        1        2
8:      1       0        2        1
9:      1       0        3        6
10:     1       1        1        3
11:     1       1        2        2
12:     1       1        3        4

> grade  <- c(0,0,0,0,0,0,1,1,1,1,1,1)
> gender <- c(0,0,0,1,1,1,0,0,0,1,1,1)
> type   <- c(1,2,3,1,2,3,1,2,3,1,2,3)
> count  <- c(3,4,2,2,4,3,2,1,6,3,2,4)

> mydata2 <-data.frame(grade, gender, type, count)
> mymod3 <- glm(grade ~ gender + factor(type),
                weights=count,
                family=binomial,
                data=mydata2)
> summary(mymod3)
                         .    .    .
Coefficients:
               Estimate Std. Error z value Pr(>|z|)
(Intercept)    -0.04941    0.72517  -0.068    0.946
gender          0.09883    0.70889   0.139    0.889
factor(type)2 -0.98587     0.92758  -1.063    0.288
factor(type)3  0.69685     0.83742   0.832    0.405

    Null deviance: 49.907  on 11  degrees of freedom
Residual deviance: 45.830  on  8  degrees of freedom
AIC: 53.83

> source("c://Rfiles/toOR.R")
> toOR(mymod3)
                  or  delta  zscore  pvalue exp.loci. exp.upci.
(Intercept)   0.9518 0.6902 -0.0681 0.9457 0.2298     3.9428
gender        1.1039 0.7825  0.1394 0.8891 0.2751     4.4292
factor(type)2 0.3731 0.3461 -1.0628 0.2879 0.0606     2.2983
factor(type)3 2.0074 1.6811  0.8321 0.4053 0.3889    10.3622
```

This output is a complete logistic model of the table. Predicted values and residuals as defined in the section on residuals in this chapter can be used to further evaluate the model. As it exists, though, the model is a poor one. However, other table data can lead to a model that is well fitted and meaningful.

SAS CODE

```
/* Section 4.1 */

*Build the logistic model and obtain the Person residual;
proc genmod data=medpar descending;
        class type (ref='1') / param=ref;
        model died=white hmo los type / dist=binomial link=logit;
        output out=residuals reschi=pearson;
run;

*Pearson Chi2 statistic;
proc sql;
        create table pr as
        select sum(pearson**2) as pchi2, 1489 as df,
          1-probchi(sum(pearson**2), 1489) as pvalue
        from residuals;
quit;

*Refer to proc print in section 2.2 to print dataset pr-Chi2 statistic;

*Type3 option provides the likelihood ratio test;
proc genmod data=medpar descending;
        class type (ref='1') / param=ref;
        model died=white hmo los type / dist=binomial link=logit type3;
run;

*Anscombe residuals can be obtained as a model output in the SAS/Insight,
not in SAS command language;

*Create new variables;
data mylgg;
set medpar;
if died=1 then dead=1;
else if died=0 then alive=1;
drop died;
m=sum(alive, dead);
run;

*Transform the dataset;
proc sql;
        create table mylgg1 as
        select white as white, hmo as hmo, type as type, count(alive) as
        alive, count(dead) as dead, count(m) as m
        from mylgg
        group by white, hmo, type;
quit;


*Obstats option provides all the residuals and statistics in Table 4.2;
proc genmod data=medpar descending;
        class type (ref='1') / param=ref;
        model died=white hmo los type / dist=binomial link=logit obstats;
        ods output obstats=stats;
```

```
run;

*Square the standardized deviance residual;
data stats1;
        set stats;
        stresdev2=stresdev**2;
run;

*Plot the square of standardized deviance residuals and mu;
proc gplot data=stats1;
        symbol v=circle color=black;
        plot stresdev2*pred / vref=4 cvref=red;
run;

*Plot the leverage and std Pearson residual;
proc gplot data=stats1;
        symbol v=circle color=black;
        plot leverage*streschi / href=0 chref=red;
run;

*Sort the dataset;
proc sort data=medpar out=medpar1;
        by white hmo los type;
run;

*Calculate the sum of the dead;
proc means data=medpar1 sum;
        by white hmo los type;
        var died;
        output out=summary sum=dead;
run;

*Create a new variable alive;
data summary1;
        set summary;
        alive=_freq_-dead;
        drop _type_ _freq_;
run;

*Refer to proc print in section 2.2 to print dataset summary1;

*Build the logistic model with numeric variables;
proc genmod data=medpar descending;
        model died=los type/dist=binomial link=logit;
run;

*Output the los;
proc freq data=medpar;
        tables los/out=los;
run;

*Prepare for the conditional effects plot;
data effect;
        set los;
        k1=-0.8714+(-0.0376)*los+0.4386*1;
        r1=1/(1+exp(-k1));
```

```
        k2=-0.8714+(-0.0376)*los+0.4386*2;
        r2=1/(1+exp(-k2));
        k3=-0.8714+(-0.0376)*los+0.4386*3;
        r3=1/(1+exp(-k3));
run;

*Graph the conditional effects plot;
proc sgplot data=effect;
        scatter x=los y=r1;
        scatter x=los y=r2;
        scatter x=los y=r3;
        xaxis label='LOS';
        yaxis label='Type of Admission' grid values=(0 to 0.4 by 0.1);
        title 'P[Death] within 48 hr admission';
run;

/* Section 4.2 */

*Build the logistic model and output model prediction;
proc genmod data=medpar descending;
        class type (ref='1') / param=ref;
        model died=white hmo los type / dist=binomial link=logit;
        output out=fit pred=mu;
run;

*Refer to proc means in section 2.5 to calculate the mean;

*Build the logistic model and output classification table & ROC curve;
proc logistic data=medpar descending plots(only)=ROC;
        class type (ref='1') / param=ref;
        model died=white hmo los type / outroc=ROCdata ctable pprob=(0 to
           1 by 0.0025);
        ods output classification=ctable;
run;

*Sensitivity and specificity plot;
symbol1 interpol=join color=vibg height=0.1 width=2;
symbol2 interpol=join color=depk height=0.1 width=2;
axis1 label=("Probability") order=(0 to 1 by 0.25);
axis2 label=(angle=90 "Sensitivity Specificity %") order=(0 to 100 by 25);
proc gplot data=ctable;
        plot sensitivity*problevel specificity*problevel /
        overlay haxis=axis1 vaxis=axis2 legend;
run;

*Approximate cutoff point can be found when sensitivity and specificity
are closest/equal in the classification table;

/* Section 4.3 */

*Lackfit option provides the Hosmer-Lemeshow GOF test;
proc logistic data=medpar descending;
        class type (ref='1') / param=ref;
        model died=white hmo los type / lackfit;
run;

/* Section 4.4 */
```

```
*Refer to the code in section 1.4 to import HIV dataset;

*Build the weighted logistic model;
proc genmod data=HIV descending;
        class cd4 (ref='0') cd8 (ref='0') / param = ref;
        weight cases;
        model infec= cd4 cd8 / dist=binomial link=logit;
run;

*Build the Firth's penalized logistic model;
proc logistic data=HIV descending;
        class cd4 (ref='0') cd8 (ref='0') / param = ref;
        weight cases;
        model infec= cd4 cd8 / firth clodds=pl;
run;

/* Section 4.5 */

*Refer to the code in section 1.4 to import and print azheart dataset;

*Generate a table of died by procedure and type;
proc freq data=azheart;
        tables died*procedure died*type / norow nocol nocum nopercent;
run;

*Build the logistic model and obtain odds ratio & statistics;
proc genmod data=azheart descending;
        model died=procedure type / dist=binomial link=logit;
        estimate "Intercept" Intercept 1 / exp;
        estimate "Procedure" procedure 1 / exp;
        estimate "Type" type 1 / exp;
run;

*Build the quasibinomial logistic model;
proc glimmix data=azheart;
        model died (event='1')=procedure type/dist=binary link=logit
         solution covb;
        random _RESIDUAL_;
run;

*Refer to proc iml in section 2.3 and the full code is provided online;

*Build the exact logistic model;
proc genmod data=azheart descending;
        model died=procedure type / dist=binomial link=logit;
        exact procedure type / estimate=both;
run;

/* Section 4.6 */

*Refer to data step in section 2.1 if manually input mydata dataset;

*Build the logistic model with weight and obtain odds ratio;
proc genmod data=mydata descending;
        weight count;
```

```
        model y=x / dist=binomial link=logit;
        estimate "Intercept" Intercept 1 / exp;
        estimate "x" x 1 / exp;
run;

*Refer to data step in section 2.1 if manually input mydata2 dataset;

*Build the logistic model with weight and obtain odds ratio;
proc genmod data=mydata2 descending;
        class type (ref='1') / param=ref;
        weight count;
        model grade=gender type / dist=binomial link=logit;
        estimate "Intercept" Intercept 1 / exp;
        estimate "Gender" gender 1 / exp;
        estimate "Type2" type 1 0 / exp;
        estimate "Type3" type 0 1 / exp;
run;
```

STATA CODE

```
4.1
. use medpar
. xi: logit died white los hmo i.type, nolog
. lrdrop1
. qui logit died white hmo los i.type, nolog
. estimates store A
. qui logit died white hmo los, nolog
. estimates store B
. lrtest A B
. predict mu
. gen died - mu                     # raw residual
. predict dev, deviance             # deviance resid
. predict pear, residuals           # Pearson resid
. predict, hat, hat                 # hat matrix diagonal
. gen stddev = dev/sqrt(1-hat)      # standardized deviance
. predict, stpear, rstandard        # standardized Pearson
. predict deltadev, ddeviance       # delta deviance
. predict dx2, dx2                  # delta Pearson
. predict dbeta, dbeta              # delta beta
. scatter stdev^2 mu
. scatter hat stpear
. qui glm died los admit, fam(bin)
. gen L1= _b[_cons] + _b[los]*los + _b[admit]*1  # Cond. effects plot
. gen Y1 = 1/(1+exp(-L1))
. gen L2 = _b[_cons] + _b[los]*los + _b[admit]*0
. gen Y1 = 1/(1+exp(-L2))
. scatter Y1 Y2 age, title("Prob of death w/I 48 hrs by admit type")

4.2
. glm died white hmo los i.type, fam(bin)
. predict mu
```

```
. mean died
. logit died white hmo los i.type, nolog
. lsens, genprob(cut) gensens(sen) genspec(spec)
. lroc
. estat classification, cut(.351)
```

4.3

```
. estat gof, tabls cut(10)
. estat gof, tabls cut(12)
```

4.4

```
. use hivlgold
. list
. glm infec i.cd4 i.cd8 [fw=cases], fam(bin)
. firthlogit infec i.cd4 i.cd8 [fw=cases], nolog
```

4.5

```
. use azcabgptca34
. list in 1/6
. table died procedure
. table died type
. glm died procedure type, fam(bin) nolog
. glm died procedure type,fam(bin) scale(x2) nolog
. exlogistic died procedure type, nolog
```

4.6

```
. use pgmydata
. glm y x [fw=count], fam(bin) nolog
. glm y x [fw=count], fam(bin) nolog eform
. use phmydata2
. glm grade gender i.type [fw=count], fam9bin) nolog nohead
. glm grade gender i.type [fw=count], fam9bin) nolog nohead eform
```

Grouped Logistic Regression

5

5.1 THE BINOMIAL PROBABILITY DISTRIBUTION FUNCTION

Grouped logistic regression is based on the binomial probability distribution. Recall that standard logistic regression is based on the Bernoulli distribution, which is a subset of the binomial. As such, the standard logistic model is a subset of the grouped. The key concept involved is the binomial probability distribution function (PDF), which is defined as:

$$f(y;p,n) = \binom{n}{y} p^y (1-p)^{n-y} \tag{5.1}$$

The product sign is assumed to be in front of the right-hand side terms. In exponential family form, the above expression becomes:

$$f(y;p,n) = \exp\left\{ y \ln\left(\frac{p}{1-p}\right) + n \ln(1-p) + \binom{n}{y} \right\} \tag{5.2}$$

The symbol n represents the number of observations in a given covariate pattern. We have discussed covariate patterns before when dealing with residuals in the last chapter. For Bernoulli response logistic models, the model is estimated on the basis of observations. Only when analyzing the fit of the model is the data put into covariate patterns and evaluated by observation-based residuals. Here the PDF itself is in covariate pattern structure.

The first derivative of the cumulant, $-n \ln(1-p)$, with respect to the link, $\ln(p/(1-p))$, is the mean, which for the binomial distribution is

$$\text{Mean} = \mu = np$$

and the second derivative of the cumulant with respect to the link is the variance.

$$\text{Variance} = V(Y) = np(1-p)$$

or, in terms of μ

$$\text{Variance} = V(\mu) = \mu\left(1 - \frac{\mu}{n}\right) = n\mu(n-\mu)$$

The link function in terms of μ is

$$\text{Link} = \ln\left(\frac{\mu}{n-\mu}\right)$$

The inverse link, which defines μ in terms of η, or xb, is

$$\text{Inverse link} = \frac{n}{1+\exp(-xb)} = \frac{n\exp(xb)}{1+\exp(xb)}$$

The log-likelihood function, with subscripts indicating individual observations

$$\mathcal{L}(\mu_i; y_i, n_i) = \sum_{i=1}^{m}\left\{y_i \ln\left(\frac{\mu_i}{1-\mu_i}\right) + n_i \ln(1-\mu_i) + \binom{n_i}{y_i}\right\} \tag{5.3}$$

Finally, the deviance statistic is defined as:

$$D = 2\sum_{i=1}^{m}\left\{y_i \ln\left(\frac{y_i}{\mu_i}\right) + (n_i - y_i)\ln\left(\frac{n_i - y_i}{n_i - \mu_i}\right)\right\} \tag{5.4}$$

5.2 FROM OBSERVATION TO GROUPED DATA

Many data sets we have to model are structured in the following format:

```
y   cases x1   x2   x3
1   3     1    0    1
1   1     1    1    1
2   2     0    0    1
0   1     0    1    1
2   2     1    0    0
0   1     0    1    0
```

*x*1, *x*2, and *x*3 are all binary predictors. The variable *cases* have values that inform us of the number of times these three binary predictors have the same values—if the data were in observation format. *y* indicates how many of the number of cases with the same covariate pattern have 1 as a value for *y*. The first line represents three observations having $x1 = 1$, $x2 = 0$, and $x3 = 1$. One of the three observations has $y = 1$, and two have $y = 0$. In observation format the above grouped data set appears as

```
y     x1     x2     x3          Line from grouped data above
1     1      0      1                     1
0     1      0      1                     1
0     1      0      1                     1
1     1      1      1                     2
1     0      0      1                     3
1     0      0      1                     3
0     0      1      1                     4
1     1      0      0                     5
1     1      0      0                     5
0     0      1      0                     6
```

This data set is in observation-based form, with *y* as 0 or 1. But you should be able to clearly see that both of the data sets are identical, providing exactly the same information. We shall model both to be sure. To do so, both must be put into separate data frames.

Observation Data

```
> y  <- c(1,0,0,1,1,1,0,1,1,0)
> x1 <- c(1,1,1,1,0,0,0,1,1,0)
> x2 <- c(0,0,0,1,0,0,1,0,0,1)
```

```
> x3 <- c(1,1,1,1,1,1,1,0,0,0)
> obser <- data.frame(y,x1,x2,x3)
> xx1 <- glm(y ~ x1 + x2 + x3, family = binomial, data = obser)
> summary(xx1)

          .    .    .

Coefficients:
            Estimate Std. Error z value Pr(>|z|)
(Intercept)   1.2050     1.8348   0.657    0.511
x1            0.1714     1.4909   0.115    0.908
x2           -1.5972     1.6011  -0.998    0.318
x3           -0.5499     1.5817  -0.348    0.728

    Null deviance: 13.46 on 9 degrees of freedom
Residual deviance: 12.05 on 6 degrees of freedom
AIC: 20.05
```

Grouped Data

```
> y     <- c(1,1,2,0,2,0)
> cases <- c(3,1,2,1,2,1)
> x1    <- c(1,1,0,0,1,0)
> x2    <- c(0,1,0,1,0,1)
> x3    <- c(1,1,1,1,0,0)
> grp   <- data.frame(y,cases,x1,x2,x3)
> grp$noty <- grp$cases - grp$y
> xx2 <- glm( cbind(y, noty) ~ x1 + x2 + x3, family = binomial, data = grp)
> summary(xx2)

Coefficients:
            Estimate Std. Error z value Pr(>|z|)
(Intercept)   1.2050     1.8348   0.657    0.511
x1            0.1714     1.4909   0.115    0.908
x2           -1.5972     1.6011  -0.998    0.318
x3           -0.5499     1.5817  -0.348    0.728

(Dispersion parameter for binomial family taken to be 1)

    Null deviance: 9.6411 on 5 degrees of freedom
Residual deviance: 8.2310 on 2 degrees of freedom
AIC: 17.853
```

The coefficients, standard errors, z values, and p-values are identical. However, the ancillary deviance and AIC statistics differ due to the number of observations in each model. But the information in the two data sets is the same. This point is important to remember.

Note that the response variable is *cbind(y, noty)* rather than *y* as in the standard model. R users tend to prefer having the response be formatted in terms of two columns of data—one for the number of 1s for a given covariate pattern, and the second for the number of 0s (not 1s). It is the only logistic regression software I know of that allows this manner of formatting the binomial response. However, one can create a variable representing the *cbind(y, noty)* and run it as a single term response. The results will be identical.

```
> grp2 <- cbind(grp$y, grp$noty)
> summary(xx3 <- glm( grp2 ~ x1 + x2 + x3, family = binomial, data = grp))

          . . .
Coefficients:
            Estimate Std. Error z value Pr(>|z|)
(Intercept)   1.2050     1.8348   0.657    0.511
x1            0.1714     1.4909   0.115    0.908
x2           -1.5972     1.6011  -0.998    0.318
x3           -0.5499     1.5817  -0.348    0.728
```

In a manner more similar to that used in other statistical packages, the binomial denominator, *cases*, may be employed directly into the response—but only if it is also used as a weighting variable. The following code produces the same output as above,

```
> summary(xx4 <- glm( y/cases ~ x1 + x2 + x3, family = binomial,
                              weights = cases, data = grp))
          . . .
Coefficients:
            Estimate Std. Error z value Pr(>|z|)
(Intercept)   1.2050     1.8348   0.657    0.511
x1            0.1714     1.4909   0.115    0.908
x2           -1.5972     1.6011  -0.998    0.318
x3           -0.5499     1.5817  -0.348    0.728
```

The advantage of using this method is that the analyst does not have to create the *noty* variable. The downside is that some postestimation functions do not accept being based on a weighted model. Be aware that there are alternatives and use the one that works best for your purposes. The *cbind()* response appears to be the most popular, and seems to be used more in published research.

Stata and SAS use the grouping variable; for example, *cases*, as the variable n in the binomial formulae listed in the last section and as given in the example directly above. The binomial response can be thought of as y = numerator and *cases* = denominator. Of course these term names will differ depending on the data being modeled. Check the end of this chapter for how Stata and SAS handle the binomial denominator.

At times a data set may be too large to simply transcribe an observation to grouped format. We will see later in this chapter why converting a categorical observation logistic model to a grouped model is desirable. In any case, using the code discussed in Chapter 4, Section 4.1.3, we may convert the above observation data set, *obser*, to a *cbind*()-based grouped format and run. I will show the data again for clarity.

```
> y  <- c(1,0,0,1,1,1,0,1,1,0)
> x1 <- c(1,1,1,1,0,0,0,1,1,0)
> x2 <- c(0,0,0,1,0,0,1,0,0,1)
> x3 <- c(1,1,1,1,1,1,1,0,0,0)
> obser <- data.frame(y,x1,x2,x3)
> xx1 <- glm(y ~ x1 + x2 + x3, family = binomial, data = obser)
```

```
> library(reshape)
> obser$x1 <- factor(obser$x1)
> obser$x2 <- factor(obser$x2)
> obser$x3 <- factor(obser$x3)
> grp <- na.omit(data.frame(cast(melt(obser, measure = "y"),
    x1 + x2 + x3 ~ .,
    function(x) { c(notyg = sum(x==0), yg = sum(x==1)) } )))
> grp
  x1 x2 x3 notyg yg
1  0  0  1     0  2
2  0  1  0     1  0
3  0  1  1     1  0
4  1  0  0     0  2
5  1  0  1     2  1
6  1  1  1     0  1
```

```
> bin <- glm( cbind(notyg, yg) ~ x1+x2+x3, famil=binomial, data=grp)
> summary(bin)
  . . .
Coefficients:
            Estimate Std. Error z value Pr(>|z|)
(Intercept)  -1.2050     1.8348  -0.657    0.511
x11          -0.1714     1.4909  -0.115    0.908
x21           1.5972     1.6011   0.998    0.318
x31           0.5499     1.5817   0.348    0.728

                  .     .     .
```

The code used to convert an observation to a grouped data set is the same code that can convert an *n*-asymptotic data set to *m*-symptotic for residual analysis. You can use the above code as a paradigm for converting any observation data to grouped format.

5.3 IDENTIFYING AND ADJUSTING FOR EXTRA DISPERSION

Grouped logistic regression models can have more correlation in the data than is allowed on the basis of binomial distributional assumptions. If data is gathered in panels or is clustered, there is likely more correlation within clusters than between them. This fact violates the criterion of independence of observations that is required of probability functions. In such a situation the data is commonly said to be overdispersed.

It is often written that a binary response logistic model cannot be overdispersed because it lacks a separate parameter for the variance; for example, the variance parameter, σ^2, in the normal linear regression or Gaussian model. Given a specific value for the mean, μ, one directly knows the value of the variance, $\mu(1 - \mu)$. If $\mu = 0.3$, then the variance is equal to 0.21. In one sense this is definitely the case, but in another it is not. If you are familiar with count models, and the Poisson model in particular, you know that the Poisson mean and variance functions are identical, that is, mean = μ; variance = μ. Yet statisticians commonly discuss Poisson overdispersion—when there is more variability in the data than is allowed by the Poisson distributional assumption of the equality of the mean and variance. Adjustments are made to the Poisson model to adjust for overdispersion in the count data.

In a similar manner to count data, binary response data can also be correlated, leading to extra dispersion in the data. In R, the *glm* function models such binomial data with a quasibinomial "family," as it uses the quasipoisson "family" for overdispersed Poisson data. Both methods scale the model standard errors by the square root of the dispersion statistic. It is a *post hoc* method applied after estimation. The grouped logistic model adds a binomial denominator to the model; e.g., as in the *cases* variable we used in the last section. The point here is that we have seen that the data in observation format (binary response) is identical to the data in grouped format (binomial response). Modeling both give the identical coefficients and standard errors. If a binomial model is overdispersed, the observation based model must also have intrinsic extra correlation. In Hilbe (2009) I call this type of binary model correlation *implicit overdispersion*. The fact that analysts employ scaling, robust or sandwich adjustors, and so forth, to correlated binary models belies the fact that they are adjusting for extra correlation or overdispersion in the data. Remember that a Bernoulli model is a binomial or grouped model—but one with a binomial denominator of 1. This indicates that each observation is a separate denominator.

Analysts can actually create a model that specifically adds an extra parameter to the model that adjusts for the extra correlation or overdispersion in the data. For the Poisson model, the negative binomial model serves this purpose. It is a two-parameter model. The beta binomial is a two-parameter logistic model, with the extra heterogeneity parameter adjusting for extra correlation in the data. Other two- and three-parameter models have also been developed to account for Poisson and binomial overdispersion, but they need not concern us here (Hilbe, 2014). We shall discuss the beta binomial later in this chapter.

How is binomial overdispersion identified? The easiest way is by using the Pearson dispersion statistic. Let us view the dispersion statistic on the grouped binomial model we created above from observation data.

```
> source("c://rfiles/P__disp.R")
> P__disp(bin)

Pearson Chi2 = 6.630003
Dispersion   = 3.315001
```

Any value of the dispersion greater than 1 indicates extra variation in the data. That is, it indicates more variation than is allowed by the binomial PDF which underlies the model. Recall that the dispersion statistic is the Pearson statistic divided by the residual degrees of freedom, which is defined as the number of observations in the model less coefficients (predictors, intercept, extra parameters). The product of the square root of the dispersion by the standard error of each predictor in grouped logistic model produces a quasi-binomial grouped logistic model. It adjusts the standard errors of the model. Sandwich and bootstrapped standard errors may be used as well to adjust for overdispersed grouped logistic models.

A caveat should be given regarding the identification of overdispersed data. I mentioned that for grouped logistic models that a dispersion statistic greater than 1 indicates overdispersion, or unaccounted for variation in the data. However, there are times that models appear to be overdispersed, but are in fact not. A grouped logistic model dispersion statistic may be greater than 1, but the model data can itself be adjusted to eliminate the perceived overdispersion. Apparent overdispersion occurs in the following conditions:

Apparent Overdispersion

- The model is missing a needed predictor.
- The model requires one or more interactions of predictors.
- A predictor needs to be transformed to a different scale; log(x).
- The link is misspecified (the data should be modeled as probit or cloglog).
- There are existing outliers in the data.

Examples of how these indicators of apparent overdispersion affect logistic models are given in Hilbe (2009).

Guideline

> If a grouped logistic model has a dispersion statistic greater than 1, check each of the 5 indicators of apparent overdispersion to determine if applying them reduces the dispersion to approximately 1. If it does, the data are not truly overdispersed. Adjust the model accordingly. If the dispersion statistic of a grouped logistic model is less than 1, the data is under-dispersed. This type of extra-dispersion is more rare, and is usually dealt with by scaling or using robust SEs.

5.4 MODELING AND INTERPRETATION OF GROUPED LOGISTIC REGRESSION

Modeling and interpreting grouped logistic models is the same as for binary response, or observation-based models. The graphics that one develops will be a bit different from the ones developed that are based on a binary response model. Using the *mylgg* model we developed in Chapter 4, Section 4.1.3 when discussing residual analysis, we shall plot the same leverage versus standardized Pearson residuals (Figure 5.1) and standardized deviance residuals versus *mu* (Figure 5.2) as done in Chapter 4. However, this time the standardized residuals in Figure 5.2 are not squared. For a binary response model, squaring the standardized residuals provides for an easier interpretation. Note the difference due to the grouped format of the data.

```
> fit  <- glm( cbind(dead, alive) ~ white + hmo + los + factor(type),
                     family = binomial, data = mylgg)
> mu   <- fit$fitted.value                # predicted probability
> hat  <- hatvalues(fit)                  # hat matrix diagnoal
> dr   <- resid(fit, type = "deviance")   # deviance residuals
> pr   <- resid(fit, type = "pearson")    # Pearson residuals
> stdr <- dr/sqrt(1-hat)                  # standardized deviance
> stpr <- pr/sqrt(1-hat)                  # standardized Pearson

> plot(stpr, hat)                         # leverage plot
> abline(v = 0, col = "red")
```

The interpretation of the *hat* statistics is the same as in Chapter 4. In Figure 5.2, notice the more scattered nature of the standardized deviance residuals. This is due to the variety of covariate patterns. Covariate patterns higher than the line at 2 are outliers, and do not fit the model.

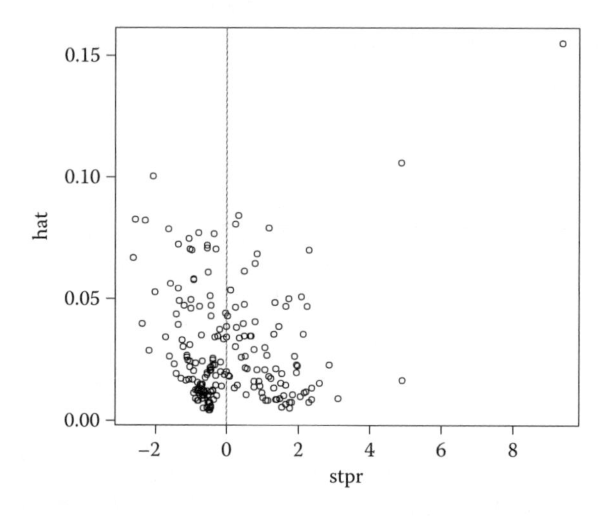

FIGURE 5.1 Leverage versus standardized Pearson.

```
plot(mu, stdr)
abline(h = 4, lty = "dotted", col = "red")
```

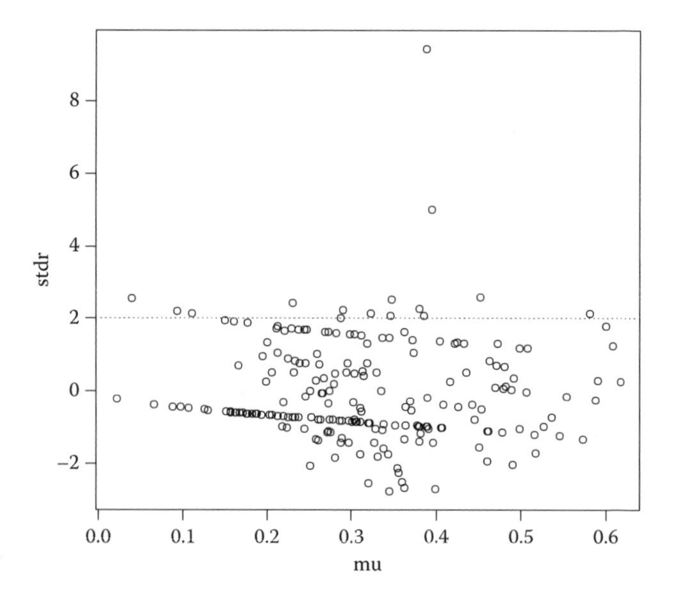

FIGURE 5.2 Standardized deviance versus *mu*.

5.5 BETA-BINOMIAL REGRESSION

Modeling overdispersed binomial data can be developed by assuming that the binomial mean parameter is itself beta distributed. That is, we provide a prior beta distribution to μ, the logistic model probability of success, or 1. The beta distribution, unlike the binomial, is a doubly bounded two-parameter distribution. This second parameter is employed in the model to adjust for any extra-binomial correlation found in the data. The two-parameter model, which is based on a mixture of beta and binomial distributions, is known as beta-binomial regression.

The binomial distribution below is expressed in terms of parameter μ. This is standard when the binomial distribution is being modeled as a generalized linear model (GLM), otherwise the parameter is typically symbolized as π. Since I will use the *glm* function in R when modeling the binomial component of the beta binomial, we shall employ μ in place of π. Moreover, I shall not use subscripts with the formulae displayed below unless otherwise noted.

Binomial PDF

$$f(y;\mu,n) = \binom{n}{y}\mu^{y}(1-\mu)^{n-y} \tag{5.5}$$

As discussed before, the $\binom{n}{y}$ choose function is the binomial coefficient, which is the normalization term of the binomial PDF. It guarantees that the function sums to 1.0. This form of the function may also be expressed in terms of factorials:

$$\binom{n}{y} = \frac{n!}{y!(n-y)!} \tag{5.6}$$

which is easily recognized from basic algebra as a combination. Both terms can be interpreted as describing the number of ways that y successes can be distributed among n trials, or observations. Note though that the mean parameter, μ, is not a term in the coefficient.

Using the Greek symbol Γ for a gamma function, $\Gamma()$, the binomial normalization term from Equation 5.6 above may be expressed as:

$$\frac{\Gamma(n+1)}{\Gamma(y+1)\Gamma(n-y+1)} \tag{5.7}$$

The log-likelihood function for the binomial model can then be expressed, with subscripts, as:

$$\mathcal{L}(\mu_i; y_i, n_i) = \sum_{i=1}^{n} \{y_i \ln(\mu_i) + (n_i - y_i)\ln(1-\mu_i) + \ln\Gamma(n_i+1) - \ln\Gamma(y_i+1) - \ln\Gamma(n_i - y_i + 1)\} \tag{5.8}$$

The beta distribution is used as the basis of modeling proportional data. That is, beta data is constrained between 0 and 1—and can be thought of in this context as the proportion obtained by dividing the binomial numerator by the denominator. The beta PDF is given below in terms of two shape parameters, a and b, although there are a number of different parameterizations.

Beta PDF

$$f(y; a, b) = \frac{\Gamma(a+b)}{\Gamma(a)\Gamma(b)} y^{a-1}(1-y)^{b-1} \tag{5.9}$$

where a is the number of successes and b is the number of failures. The initial term in the function is the normalization constant, comprised of gamma functions.

The above function can also be parameterized in terms of μ. Since we plan on having the binomial parameter, μ, itself distributed as beta, we can parameterize the beta PDF as:

$$f(\mu) = \frac{\Gamma(a+b)}{\Gamma(a)\Gamma(b)} \mu^{a-1}(1-\mu)^{b-1} \tag{5.10}$$

Notice that the kernal of the beta distribution is similar to that of the binomial kernal.

$$\mu^{y}(1-\mu)^{n-y} \sim \mu^{a-1}(1-\mu)^{b-1} \tag{5.11}$$

Even the coefficients of the beta and binomial are similar in structure. In probability theory such a relationship is termed *conjugate*. The beta distribution is conjugate to the binomial. This is a very useful property when mixing distributions, since it generally allows for easier estimation. Conjugacy plays a particularly important role in Bayesian modeling where a prior conjugate (beta) distribution of a model coefficient, which is considered to be a random variable, is mixed with the (binomial) likelihood to form a beta posterior distribution.

The mean and variance of the beta PDF may be given as:

$$E(y) = \frac{a}{a+b} = \mu \quad V(y) = \frac{ab}{(a+b)^2(a+b+1)} \tag{5.12}$$

As mentioned before, the beta-binomial distribution is a mixture of the binomial and beta distributions. The binomial parameter, μ, is distributed as beta, which adjusts for extra-binomial correlation in the data. The mixture can be obtained by multiplying the two distributions.

$$f(y;\mu,a,b) = f(y;\mu,n)f(y;\mu,a,b) \tag{5.13}$$

The result is the beta-binomial probability distribution.

Beta Binomial

$$f(y;\mu,a,b) = \frac{\Gamma(a+b)\Gamma(n+1)}{\Gamma(a)\Gamma(b)\Gamma(y+1)\Gamma(n-y+1)}\pi^{y-a-1}(1-\mu)^{n-y+b-1} \tag{5.14}$$

An alternative parameterization may be given in terms of μ and σ, with $\mu = a/(a+b)$.

$$f(y;\mu,\sigma) = \frac{\Gamma(n+1)}{\Gamma(y+1)\Gamma(n-y+1)} \frac{\Gamma\left(\frac{1}{\sigma}\right)\Gamma\left(y+\frac{\mu}{\sigma}\right)\Gamma\left(n-y+\frac{1-\mu}{\sigma}\right)}{\Gamma\left(n+\frac{1}{\sigma}\right)\Gamma\left(\frac{\mu}{\sigma}\right)\Gamma\left(\frac{1-\mu}{\sigma}\right)} \tag{5.15}$$

with $y = 0, 1, 2, \ldots n$, and $0 < \mu < 1$, and $\sigma > 0$.

Under this parameterization, the mean and variance of the beta binomial are:

$$E(Y) = n\mu \quad V(Y) = n\mu(1-\mu)\left[1 + \frac{\sigma}{1+\sigma}(n-1)\right] \tag{5.16}$$

This is the parameterization that is used in R's *gamlss* function (Rigby and Stasinopoulos, 2005) and in the Stata *betabin* command (Hardin and Hilbe, 2014).

For an example, we shall use the 1912 Titanic shipping disaster passenger data. In grouped format, the data are called *titanicgrp*. The predictors of the model are:

Age: 1 = adult; 0 = child
Sex: 1 = male; 0 = female
Class: 1st class, 2nd class, 3rd class (we make 3rd class the reference)

The response is how many survived given a specific covariate pattern. *Cases* represents the number of passengers having the same predictor values.

```
> data(titanicgrp)
> titanicgrp ; attach(titanicgrp) ; table(class)
   survive cases    age   sex      class
1        1     1  child women 1st class
2       13    13  child women 2nd class
3       14    31  child women 3rd class
4        5     5  child   man 1st class
5       11    11  child   man 2nd class
6       13    48  child   man 3rd class
7      140   144 adults women 1st class
8       80    93 adults women 2nd class
9       76   165 adults women 3rd class
10      57   175 adults   man 1st class
11      14   168 adults   man 2nd class
12      75   462 adults   man 3rd class
```

Change the default reference to 3rd class

```
> class03 <- factor(titanicgrp$class,
       levels=c("3rd class", "2nd class", "1st class"))
```

Set up and run the grouped logistic model

```
> died <- titanicgrp$cases - titanicgrp$survive
> summary(jhlogit <- glm(cbind(survive,died) ~ age+sex+class03,
                   data=titanicgrp, family=binomial))
Coefficients:
                Estimate Std. Error z value Pr(>|z|)
(Intercept)       1.2955     0.2478   5.227 1.72e-07 ***
ageadults        -1.0556     0.2427  -4.350 1.36e-05 ***
sexman           -2.3695     0.1453 -16.313  < 2e-16 ***
class032nd class  0.7558     0.1753   4.313 1.61e-05 ***
class031st class  1.7664     0.1707  10.347  < 2e-16 ***
---

 Null deviance: 581.40 on 11 degrees of freedom
Residual deviance: 110.84 on 7 degrees of freedom
AIC: 157.77

> source("c://rfiles/toOR.R") # or from LOGIT package
> toOR(jhlogit)
                  or  delta  zscore pvalue exp.loci. exp.upci.
(Intercept)   3.6529 0.9053  5.2271      0    2.2473    5.9374
```

```
ageadults          0.3480 0.0844  -4.3502     0     0.2163     0.5599
sexman             0.0935 0.0136 -16.3129     0     0.0704     0.1243
class032nd class 2.1293 0.3732   4.3126     0     1.5103     3.0021
class031st class 5.8496 0.9986  10.3468     0     4.1861     8.1741

> P__disp(jhlogit)

Pearson Chi2 = 100.8828
Dispersion   = 14.41183
```

All of the predictors appear to significantly contribute to the understanding of passenger survival. The model is severely overdispersed as evidenced by a dispersion statistic of 14.4.

Next we create sandwich or robust adjustments of the standard errors. This will adjust for much of the excess correlation in the data. But the dispersion is very high.

```
> library(sandwich)
> or <- exp(coef(jhlogit))
> rse <- sqrt(diag(vcovHC(jhlogit, type = "HC0"))) # robust SEs
> ORrse <- or*rse
> pvalue <- 2*pnorm(abs(or/ORrse), lower.tail = FALSE)
> rotab <- data.frame(or, ORrse, pvalue)
> rotab
                         or      ORrse     pvalue
(Intercept)      3.65285874 2.78134854 0.18906811
ageadults        0.34798085 0.24824238 0.16098137
sexman           0.09353076 0.04414139 0.03409974
class032nd class 2.12934342 1.26408082 0.09208519
class031st class 5.84958983 3.05415838 0.05545591
```

The robust p-values tell us that *age* and *2nd class* are not significant. 1st class passengers is marginal, but given the variability in the data we would keep it in a final model, with a combined 2nd and 3rd class as the reference. That is, it may be preferred to dichotomize class as a binary predictor with 1 = 1st class and 0 = otherwise.

R output for the beta-binomial model using *gamlss* is given as displayed below. Note again that there is a slight difference in estimates. *Sigma* is the dispersion parameter, and can itself be parameterized, having predictors like the mean or location parameter, *mu*. The dispersion estimates inform the analyst which predictors significantly influence the extra correlation in the data, therefore influencing the value of *sigma*. In this form below it is only the intercept of *sigma* that is displayed. In this respect, the beta binomial is analogous to the heterogeneous negative binomial count model (Hilbe, 2011, 2014), and the binomial logistic regression function is analogous to a Poisson, or perhaps a negative binomial model.

Beta Binomial

```
> library(gamlss)
> summary(mybb <- gamlss(cbind(survive,died) ~ age + sex + class03,
          data = titanicgrp, family = BB))

                   Estimate   Std. Error  t value   Pr(>|t|)
(Intercept)           1.498       0.6814    2.199   0.063855
ageadults            -2.202       0.8205   -2.684   0.031375
sexman               -2.177       0.6137   -3.547   0.009377
class032nd class      2.018       0.8222    2.455   0.043800
class031st class      2.760       0.8558    3.225   0.014547
-------------------------------------------------------------

Sigma link function: log
Sigma Coefficients:
             Estimate  Std. Error   t value    Pr(>|t|)
(Intercept)    -1.801      0.7508    -2.399     0.03528
-------------------------------------------------------------

                   .      .      .

Global Deviance:     73.80329
Global Deviance:     73.80329
            SBC:     88.71273
```

Notice that the AIC statistic is reduced from 157.77 for the grouped logistic model to 85.80 for the beta-binomial model. This is a substantial improvement in model fit. The heterogeneity or dispersion parameter, *sigma*, is 0.165.

```
Sigma [gamlss's sigma is log(sigma)]
> exp(-1.801)
[1] 0.1651337
```

Odds ratio for beta binomial are inflated compared to the grouped logit, but the *p*-values are closely the same.

```
> exp(coef(mybb))
     (Intercept)    ageadults      sexman class032nd class
       4.4738797    0.1105858   0.1133972        7.5253615
class031st class
      15.8044343
```

I also calculated robust or sandwich standard errors for the beta-binomial model. *2nd class* and *age* resulted in nonsignificant *p*-values. This result is the same as given with the above robust grouped logit model. *gamlss* does not work well with sandwich estimators; the calculations were done using Stata. See the book's web site for results.

The beta-binomial model is preferred to the single parameter logistic model. However, extra correlation still needs to be checked and adjusted. We should check for an interactive effect between *age* and *sex*, and between both *age* and sex and *class 1*. I shall leave that as an exercise for the reader. It appears, though, from looking at the model main effects only, that females holding 1st and 2nd class tickets stood the best odds of survival on the Titanic. If they were female children, they stood even better odds. 3rd class ticket holders, and in particular 3rd class male passengers fared the worst. It should be noted that 1st class rooms were very expensive, with the best going for some US$100,000 in 2015 equivalent purchasing power.

The beta binomial is an important model, and should be considered for all overdispersed logistic models. In addition, for binomial models with *probit* and *complementary loglog* links, or with excess zero response values, Stata's *betabin* and *zibbin* commands (Hardin and Hilbe, 2013) have options for these models. Perhaps these capabilities will be made available to R users in the near future. The generalized binomial model is another function suitable for modeling overdispersed grouped logistic models. The model is available in Stata (Hardin and Hilbe, 2007) and SAS (Morel and Neerchal, 2012).

SAS CODE

```
/* Section 5.2 */

*Refer to data step in section 2.1 if manually input
obser dataset;
*Build the logistic model;
proc genmod data=obser descending;
     model y=x1 x2 x3 / dist=binomial link=logit;
run;

*Refer to data step in section 2.1 if manually input grp
dataset;

*Build the logistic model;
proc genmod data=grp descending;
```

```
     model y/cases=x1 x2 x3 / dist=binomial link=logit;
run;

*Build the logistic model;
proc genmod data=grp descending;
     class x1 (ref='0') x2 (ref='0') x3 (ref='0') / param=ref;
     model y/cases=x1 x2 x3 / dist=binomial link=logit;
run;

/* Section 5.4 */

Refer to proc sort, proc means in section 4.1 to obtain a
new dataset;

*Create a new variable alive;
data summary1;
       set summary;
       alive=_freq_-dead;
       cases=_freq_;
       drop _type_ _freq_;
run;

*Obstats option provides all the residuals and useful
statistics;
proc genmod data=summary1 descending;
       class type (ref='1')/ param=ref;
       model dead/cases=white hmo los type / dist=binomial
       link=logit obstats;
       ods output obstats=allstats;
run;

*Plot the leverage and std Pearson residual;
proc gplot data=allstats;
       symbol v=circle color=black;
       plot leverage*streschi / href=0 chref=red;
run;

*Plot the standardized deviance residuals and mu;
proc gplot data=allstats;
       symbol v=circle color=black;
       plot stresdev*pred / vref=2 cvref=red;
run;

/* Section 5.5 */

*Refer to the code in section 1.4 to import and print the
dataset;
```

```
*Build the logistic model and obtain odds ratio & covariance
matrix;
proc genmod data=titanicgrp descending;
      class class (ref='3')/ param=ref;
      model survive/cases=age sex class / dist=binomial link=logit
      covb;
      estimate "Intercept" Intercept 1 / exp;
      estimate "ageadults" age 1 / exp;
      estimate "sexman" sex 1 / exp;
      estimate "class" class 1 0 / exp;
      estimate "class" class 0 1 / exp;
run;

*Build the logistic mode with robust adjustment;
proc glimmix data=titanicgrp order=data empirical=hc0;
      class class;
      model survive/cases=age sex class/dist=binomial link=logit
      solution covb;
      random _RESIDUAL_;
run;

*Refer to proc iml in section 2.3 and the full code is provided
online;

*Build the Beta binomial model;
proc fmm data=titanicgrp;
      class class;
      model survive/cases=age sex class / dist=betabinomial;
run;
```

STATA CODE

```
5.1
. use obser
. glm y x1 x2 x3, fam(bin) nolog

5.2
. use obser, clear
. glm y x1 x2 x3, fam(bin) nolog nohead
. use grp, clear
. glm y x1 x2 x3, fam(bin cases) nolog nohead
. use obser
. gen cases=1
. collapse(sum) cases (sum) yg, by(x1 x2 x3)
. glm yg x1 x2 x3, fam(bin cases) nolog nohead
```

5.4

```
. use phmylgg
. cases = dead + alive
. glm dead white hmo los i.type, fam(bin cases)
. predict mu
. predict hat, hat
. predict dev, deviance
. gen stdev = dev/sqrt(1-hat)
. predict stpr, rstandard
. scatter stpr hat
. gen stdev2 = stdev^2
. scatter stdev2 mu
```

5.5

```
. use titanicgrp
. list
. glm died age sex b3.class, fam(bin cases) nolog
. glm, eform
. glm died age sex b3.class, fam(bin cases) vce(robust) nolog
. betabin died age sex b3.class, n(cases) nolog
```

Bayesian Logistic Regression

6

6.1 A BRIEF OVERVIEW OF BAYESIAN METHODOLOGY

Bayesian methodology would likely not be recognized by the person who is regarded as the founder of the tradition. Thomas Bayes (1702–1761) was a British Presbyterian country minister and amateur mathematician who had a passing interest in what was called inverse probability. Bayes wrote a paper on the subject, but it was never submitted for publication. He died without anyone knowing of its existence. Thomas Price, a friend of Bayes, discovered the paper when going through Bayes's personal effects. Realizing its importance, he managed to have it published in the Royal Society's *Philosophical Transactions* in 1764. The method was only accepted as a curiosity and was largely forgotten until Pierre-Simon Laplace, generally recognized as the leading mathematician worldwide during this period, discovered it several decades later and began to employ its central thesis to problems of probability. However, how Bayes's inverse probability was employed during this time is quite different from how analysts currently apply it to regression modeling. For those who are interested in the origins of Bayesian thinking, and its relationship to the development of probability and statistics in general, I recommend reading Weisberg (2014) or Mcgrayne (2011).

Inverse probability is simple in theory. Suppose that we know from epidemiological records that the probability of a person having certain symptoms S given that they have disease D is 0.8. This relationship may be symbolized as $Pr(S|D) = 0.8$. However, most physicians want to know the probability of having the disease if a patient displays these symptoms, or Pr(D|S). In order to find this out additional information is typically required. The idea is that under certain conditions one may find the inverse probability of an event, usually with the addition information. The notion of additional information is key to Bayesian methodology.

There are three foremost characteristic features that distinguish Bayesian regression models from the traditional maximum likelihood models such as logistic regression. Realize though that these features are simplifications. The details are somewhat more complicated.

1. *Parameters Are Randomly Distributed:* The regression parameters to be estimated are themselves randomly distributed. In traditional, or frequentist-based, logistic regression the estimated parameters are fixed. All main effects parameter estimates are based on the same underlying PDF.
2. *Parameters May have Different Distributions:* In Bayesian logistic regression, each parameter is separate, and may be described using a different distribution.
3. *Parameter Estimates As The Means of a Distribution:* When estimating a Bayesian parameter an analyst develops a posterior distribution from the likelihood and prior distributions. The mean (or median, mode) of a posterior distribution is regarded as the beta, parameter estimate, or Bayesian coefficient of the variable.
4. *Credible Sets Used Instead of Confidence Intervals:* Equal-tailed *credible sets* are usually defined as the outer 0.025 quantiles of the posterior distribution of a Bayesian parameter. The *posterior intervals,* or *highest posterior density* (HPD) region, are used when the posterior is highly skewed or is bi- or multi-model in shape. There is a 95% probability that the credible set or posterior mean contains the true posterior mean. Confidence intervals are based on a frequency interpretation of statistics as defined in Chapter 2, Section 2.3.4.
5. *Additional or Prior Information:* The distribution used as the basis of a parameter estimate (likelihood) can be mixed with additional information—information that we know about the variable or parameter that is independent of the data being used in the model. This is called a prior distribution.

The basic formula that defines a Bayesian model is:

$$f(\theta \mid y) = \frac{f(y \mid \theta)f(\theta)}{f(y)} = \frac{f(y \mid \theta)f(\theta)}{\int f(y \mid \theta)f(\theta)\,d\theta} \tag{6.1}$$

where $f(y|\theta)$ is the likelihood function and $f(\theta)$ is the prior distribution. The denominator, $f(y)$ is the probability of y over all y. Note that the likelihood and prior distributions are multiplied together. Usually the denominator, which is the normalization term, drops out of the calculations so that the posterior distribution or a model predictor is determined by the product of its likelihood and prior. Again, each predictor can be comprised of a different posterior.

If an analyst believes that there is no meaningful outside information that bears on the predictor, a uniform prior will usually be given. When this happens the prior is not informative.

A prior having a normal distribution with a mean of 0 and very high variance will also produce a noninformative or diffuse prior. If all predictors in the model are noninformative, the maximum likelihood results will be nearly identical to the Bayesian betas. In our first examples below we will use noninformative priors.

I should mention that priors are a way to provide a posterior distribution with more information than is available in the data itself, as reflected in the likelihood function. If a prior is weak it will not provide much additional information and the posterior will not be much different than it would be with a completely noninformative prior. In addition, what may serve as an influential informative prior in a model with few observations may well be weak when applied to data with a large number of observations.

It is important to remember that priors are not specific bits of information, but are rather distributions with parameters which are combined with likelihood distributions. A major difficulty most analysts have when employing a prior in a Bayesian model is to specify the correct parameters of the prior that describe the additional information being added to the model. Again, priors may be multiplied with the log-likelihood to form a posterior for each term in the regression.

There is much more that can be discussed about Bayesian modeling, in particular Bayesian logistic modeling. But this would take us beyond the scope we set for this book. I provide the reader with several suggested books on the subject at the end of the chapter.

To see how Bayesian logistic regression works and is to be understood is best accomplished through the use of examples. I will show an example using R's *MCMCpack* package (located on CRAN) followed by the modeling of the same data using JAGS. JAGS is regarded by many in the area as one of the most powerful, if not the most powerful, Bayesian modeling package. It was developed from *WinBUGS* and *OpenBUGS* and uses much of the same notation. However, it has more built-in functions and more capabilities than do the BUGS packages. BUGS is an acronym for "Bayesian inference Using Gibbs Sampling" and is designed and marketed by the Medical Research Group out of

Cambridge University in the United Kingdom. More will be mentioned about the BUGS packages and JAGS at the start of Chapter 6, Section 6.2.2. Stata 14 was released on April 7, 2015, well after this text was written. Stata now has full Bayesian capabilities. I was able to include Stata code at the end of this chapter for Bayesian logistic models with noninformative and Cauchy priors.

6.2 EXAMPLES: BAYESIAN LOGISTIC REGRESSION

6.2.1 Bayesian Logistic Regression Using R

For an example we shall model the 1984 German health reform data, *rwm1984*. Our variable of interest is a patient's work status. If they are not working, *outwork* = 1; if they are employed or are otherwise working, *outwork* = 0. The predictors we use to understand *outwork* are:

docvis : The number of visits made to a physician during the year, from 0 to 121.
female : 1 = female; 0 = male.
kids : 1 = has children; 0 = no children.
age : *age*, from 25 to 64.

The data are first loaded and the data are renamed *R84*. We shall view the data, including other variables in the data set.

```
> library(MCMCpack)
> library(COUNT)
> data(rwm1984)
> R84 <- rwm1984

# DATA PROFILE
> head(R84)
  docvis hospvis edlevel age outwork female married kids hhninc educ self
1      1       0       3  54       0      0       1    0  3.050 15.0    0
2      0       0       1  44       1      1       1    0  3.050  9.0    0
3      0       0       1  58       1      1       0    0  1.434 11.0    0
4      7       2       1  64       0      0       0    0  1.500 10.5    0
5      6       0       3  30       1      0       0    0  2.400 13.0    0
6      9       0       3  26       1      0       0    0  1.050 13.0    0
  edlevel1 edlevel2 edlevel3 edlevel4
1        0        0        1        0
2        1        0        0        0
```

```
3        1        0        0        0
4        1        0        0        0
5        0        0        1        0
6        0        0        1        0
```

The data have 3874 observations and 17 variables.

```
> dim(R84)
[1] 3874   17
```

The response variable, *outwork*, has 1420 1s and 2454 0s, for a mean of 0.5786.

```
> table(R84$outwork)
   0    1
2454 1420
```

Other characteristics of the data to be modeled, including the centering of both continuous predictors, are given as follows:

```
# SUMMARIES OF THE TWO CONTINUOUS VARIBLES
> summary(R84$docvis)
   Min.   1st Qu.   Median     Mean    3rd Qu.      Max.
  0.000    0.000     1.000     3.163    4.000    121.000

> summary(R84$age)
   Min.   1st Qu.   Median   Mean   3rd Qu.    Max.
     25     35         44      44       54        64

# CENTER BOTH CONTINUOUS PREDICTORS
> R84$cage <- R84$age - mean(R84$age)
> R84$cdoc <- R84$docvis - mean(R84$docvis)
```

We shall first model the data based on a standard logistic regression, and then by a logistic regression with the standard errors scaled by the square root of the Pearson dispersion. The scaled logistic model, as discussed in the previous chapter, is sometimes referred to as a "quasibinomial" model. We model both to determine if there is extra variability in the data that may require adjustments. The tables of coefficients for each model are not displayed below, but are stored in *myg* and *myq*, respectively. I shall use the *toOR* function to display the odds ratios and associated statistics of both models in close proximity. The analyst should inspect the delta (SEs) values to determine if they differ from each other by much. If they do, then there is variability in the data. A scaled logistic model, or other adjusted models, should be used on the data, including a Bayesian model. Which model we use depends on what we think is the source of the extra correlation.

```
# MODEL OF LOGISTIC (g ) AND QUASIBINOMIAL (q)
> myg <- glm(outwork ~ cdoc + female + kids + cage,
                 family=binomial, data=R84)

> myq  <- glm(outwork ~ cdoc + female + kids + cage,
                 family=quasibinomial, data=R84)


# COMPARISON OF MODEL OUTPUT - ODDS RATIOS
> source("c://Rfiles/toOR.R")
> toOR(myg)
                or  delta   zscore pvalue exp.loci. exp.upci.
(Intercept) 0.1340 0.0109 -24.7916  0e+00    0.1143    0.1570
cdoc        1.0247 0.0064   3.9012  1e-04    1.0122    1.0374
female      9.5525 0.7906  27.2691  0e+00    8.1222   11.2347
kids        1.4304 0.1287   3.9792  1e-04    1.1992    1.7063
cage        1.0559 0.0044  13.0750  0e+00    1.0473    1.0645

> toOR(myq)
                or  delta   zscore pvalue exp.loci. exp.upci.
(Intercept) 0.1340 0.0113 -23.7796  0e+00    0.1135    0.1581
cdoc        1.0247 0.0067   3.7420  2e-04    1.0117    1.0379
female      9.5525 0.8242  26.1560  0e+00    8.0663   11.3126
kids        1.4304 0.1342   3.8168  1e-04    1.1902    1.7191
cage        1.0559 0.0046  12.5413  0e+00    1.0469    1.0649
```

A comparison of the standard errors of the two models shows that there is not much extra variability in the data. The standard errors are nearly the same. No adjustments need to be made to the model. However, for pedagogical sake we shall subject the data to a Bayesian logistic regression.

Recall from Chapter 3, Section 3.4.1 that the *quasibinomial* "option" in R's *glm* function produces mistaken confidence intervals. Our *toOR* function corrects this problem for odds ratios. Log the intervals to obtain correct scaled confidence intervals.

We use the *MCMCpack* package, which has the *MCMClogit* function for estimating Bayesian logistic models. The algorithms in *MCMCpack* employ a random walk version of the Metropolis-Hastings algorithm when estimating a logistic model. MCMC is an acronym for Markov Chain Monte Carlo, which is a class of sampling algorithm used to find or determine the mean, standard deviation, and quantiles of a distribution from which the data to be modeled is theoretically derived or, at least, best described. There are a variety of algorithms employed by Bayesians which are based on MCMC; for example, Metropolis-Hastings, Gibbs Sampling.

For our example I shall employ the default *multivariate normal priors* on all of the predictors. It is used because we have more than one predictor, all of which have noninformative priors. It is therefore not necessary to show them in the formula below.

```
# BAYESIAN ANALYSIS OF MODEL
> mymc <- MCMClogit(outwork ~ cdoc + female + kids + cage,
+                          burnin = 5000,
+                          mcmc= 100000,
+                          data=R84)
```

burnin is used to tell the algorithm how many of the initial samples should be discarded before beginning to construct a posterior distribution, from which the mean, standard deviation, and quantiles are derived. *mcmc* specifies how many samples are to be used in the estimation of the posterior. We discard the first 5000 iterations and keep the next 100,000.

Options many times used in the model are `b0` and `B0`, which represent the mean and precision of the prior(s). The precision is defined as the inverse of the variance. As such one typically sees `B0` as $B0^{-1}$. Since we used the default prior of `b0 = 0` and `B0 = 0` here, assigning values to `b0` and `B0` was not required. We could have used b0 = 0 and B0 = 0.00001 as well, for a mean of 0 and an extremely high variance, which means that nothing specific is being added to the model. The priors are *noninformative*, and therefore do not appreciatively influence the model. That is, the data, or rather likelihood, is the prime influence on the parameter estimates, not the priors. An analyst may also use the `user.prior.density` option to define their own priors.

The output is given as usual:

```
> summary(mymc)

Iterations = 5001:105000
Thinning interval = 1
Number of chains = 1
Sample size per chain = 1e+05

1. Empirical mean and standard deviation for each variable,
   plus standard error of the mean:

                 Mean       SD  Naive SE Time-series SE
(Intercept) -2.01308 0.080516 2.546e-04      1.021e-03
cdoc         0.02464 0.006246 1.975e-05      7.942e-05
female       2.25923 0.083230 2.632e-04      1.073e-03
kids         0.35749 0.089348 2.825e-04      1.139e-03
cage         0.05444 0.004159 1.315e-05      5.334e-05
```

```
2. Quantiles for each variable:

                 2.5%      25%      50%      75%    97.5%
(Intercept) -2.17202 -2.06720 -2.01287 -1.95843 -1.85674
cdoc         0.01267  0.02034  0.02454  0.02883  0.03704
female       2.09550  2.20357  2.25912  2.31470  2.42444
kids         0.18391  0.29680  0.35791  0.41714  0.53193
cage         0.04630  0.05164  0.05442  0.05723  0.06255
```

Compare the output above for the noninformative prior with SAS output on the same data and model. The results are remarkably similar.

POSTERIOR SUMMARIES

PARAMETER	*N*	*MEAN*	*STANDARD DEVIATION*	*PERCENTILES* 25%	50%	75%
Intercept	100,000	−2.0140	0.0815	−2.0686	−2.0134	−1.9586
Cdoc	100,000	0.0247	0.00632	0.0204	0.0246	0.0289
Female	100,000	2.2605	0.0832	2.2043	2.2602	2.3166
Kids	100,000	0.3596	0.0907	0.2981	0.3590	0.4207
Cage	100,000	0.0545	0.00418	0.0516	0.0545	0.0573

POSTERIOR INTERVALS

PARAMETER	*ALPHA*	*EQUAL-TAIL INTERVAL*		*HPD INTERVAL*	
Intercept	0.050	−2.1755	−1.8557	−2.1710	−1.8520
Cdoc	0.050	0.0124	0.0373	0.0124	0.0373
Female	0.050	2.0989	2.4242	2.0971	2.4220
Kids	0.050	0.1831	0.5382	0.1838	0.5386
Cage	0.050	0.0463	0.0628	0.0464	0.0628

The mean values are analogous to maximum likelihood coefficients, the standard errors are like standard errors and the 2.5% and 97.5% quantiles are somewhat similar to confidence intervals. Here Bayesians refer to the external quantiles as "credible sets" or sometimes as either "credible intervals" or "posterior intervals."

Remember that each predictor is considered to be randomly distributed, and not fixed as is assumed when data are being modeled using standard frequentist-based maximum likelihood methods. As such Bayesians attempt to develop a distribution for each predictor, the mean of each is the Bayesian logistic beta. The plots on the right-hand side of Figure 6.1 display the distributions

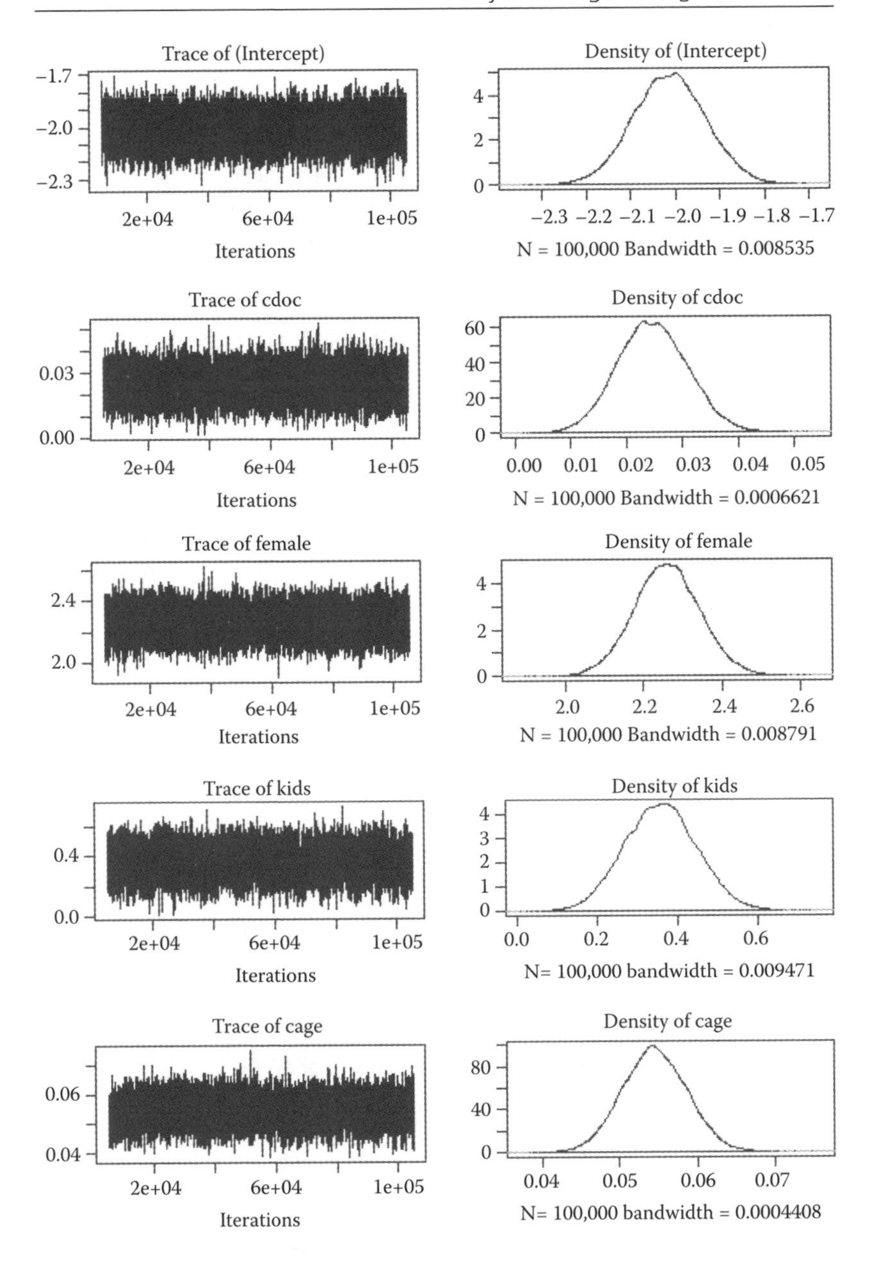

FIGURE 6.1 R trace and density plots of model with noninformative priors.

of each predictor in the model. The peak of each distribution is at the point which defines the predictor's mean. The intercept therefore is about –2.0, the mean for centered *docvis* (*cdoc*) is about 0.025, and for centered *age* (*cage*) at about 0.055. The trace plots on the left side of Figure 6.1 show time series plots across all iterations. We are looking for the convergence of the estimation to a single value. When the plot converges or stabilizes without excessive up and down on the *y* axis, convergence has been achieved. There appears to be no abnormality in the sampling draws being made by the MCMC algorithm in any of the trace plots. This is what we want to observe. In addition, if there are breaks in the trace, or places where clumps are observed, we may conclude that the sampling process is not working well.

```
> plot(mymc)      # Creates Figure 6.1
```

Geweke's diagnostic test (Geweke, 1992) is a univariate test of the equality of the means of the first 10% and final 50% of the Markov chain samples from an MCMC chain which generates the posterior distribution from which Bayesian means, standard deviations and quantiles derive. Each Geweke statistic is a *z* score and is assessed based on the normal PDF. The values in the results below only provide *z* scores, not associated *p*-values. But these are easy to obtain:

```
> geweke.diag(mymc)      # Creates Figure 6.1

Fraction in 1st window = 0.1
Fraction in 2nd window = 0.5

(Intercept)       cdoc       female        kids         cage
   1.0789      -0.8122     0.1736      -0.6669      -1.3006
```

In the table below, I have created a table of standard model coefficients and Bayesian means for comparison purposes. More importantly, the second table is a comparison of the model, scaled model, and Bayesian model standard errors or standard deviations (for Bayesian models). Notice their closeness in value. The data simply have little excess correlation or unaccounted for variability.

```
> Bcoef <- round(colMeans(mymc), 5)
> Bsd <- round(apply(mymc, MARGIN=2, FUN=sd), 5)
> mygcf <- round(coef(myg), 5)
> mygsd <- round(sqrt(diag(vcov(myg))), 5)
> myqsd <- round(sqrt(diag(vcov(myq))), 5)
> myqcf <- round(coef(myq), 5)
> Bcf <- data.frame(mygcf, Bcoef)
> Bsd <- data.frame(mygsd, myqsd, Bsd)
```

```
# COEFFICIENTS COMPARISON: MODEL AND BAYESIAN
> Bcf
               mygcf     Bcoef
(Intercept) -2.01028 -2.01308
cdoc         0.02443  0.02464
female       2.25680  2.25923
kids         0.35798  0.35749
cage         0.05438  0.05444

# COMPARISON OF STANDARD ERRORS/SD: MODEL, QUASI, BAYES
> Bsd
              mygsd   myqsd     Bsd
(Intercept) 0.08109 0.08454 0.08052
cdoc        0.00626 0.00653 0.00625
female      0.08276 0.08628 0.08323
kids        0.08996 0.09379 0.08935
cage        0.00416 0.00434 0.00416
```

MCMCpack provides basic Bayesian modeling capabilities that can be easily extended to incorporate prior and more advanced models such as level 2 random effects models. If we had information about the physician names who patients saw during 1984, we could adjust for a possible physician effect by making it a random effect. Suppose physician name is stored in the variable *physician*. The code for developing a noninformative Bayesian random effects logistic regression model is simply:

```
> mymc <- MCMClogit(outwork ~ cdoc + female + kids + cage,
+                        random = ~physician,
+                        burnin = 5000,
+                        mcmc= 100000,
+                        data=R84)
```

6.2.2 Bayesian Logistic Regression Using JAGS

JAGS is a Bayesian modeling package based on Gibbs sampling. In fact, JAGS, authored by Dutch statistician Martyn Plummer, is an acronym for *Just Another Gibbs Sampler*. First released in December 2007, the package can be run as a stand-alone program, or from within R. We shall demonstrate how it can be used within R to develop a Bayesian logistic model.

The code below is adapted from code given in Zuur et al. (2013), although the original code was designed for a completely different distribution and model. The useful aspect with this code is that it can be adapted to run a number of different models. Of course, our example will be to show its use in creating a Bayesian logistic model.

First, make sure you have installed JAGS to your computer. It is freeware, as is R. JAGS is similar to WinBUGS and OpenBUGS, which can also be run as standalone packages or within the R environment. JAGS is many times preferred by those in the hard sciences like physics, astronomy, ecology, biology, and so forth since it is command-line driven, and written in C ++ for speed. WinBUGS and OpenBUGS are written in Pascal, which tends to run slower than C ++ implementations, but can be run within the standalone WinBUGS or OpenBUGS environments, which include menus, help, and so forth. The BUGS programs are more user-friendly. Both OpenBUGS and JAGS are also able to run on a variety of platforms, which is advantageous to many users. In fact, WinBUGS is no longer being developed or supported. The developers are putting all of their attention to OpenBUGS. Lastly, and what I like about it, when JAGS is run from within R, the program actually appears as if it is just another R package. I do not feel as if I am using an outside program.

To start it is necessary to have JAGS in R's path, and the *R2jags* package needs to be installed and loaded. For the first JAGS example you also should bring two functions contained in *jhbayes.R* into memory using the source function.

```
> library(R2jags)
> source("c://Rfiles/jhbayes.R") # or where you store R
  files; book's website
```

The code in Table 6.1 is specific to the model we have been working with in the previous section. However, as you can see, it is easily adaptable for other logistic models. With a change in the log-likelihood, it can also be used with other distributions and can be further amended to incorporate random effects, mixed effects, and a host of other models.

Let us walk through the code in Table 6.1. Doing so will make it much easier for you to use it for other modeling situations.

The top two lines

```
X <- model.matrix(~ cdoc + female + kids + cage,
                  data = R84)
K <- ncol(X)
```

create a matrix of predictors, *X*, from the model *R84*, and a variable, *K*, which contains the number of predictors contained in *X*. A column of 1s for the intercept is also generated by *model.matrix*().

The next code segment is *logit.data*, although we may call it anything we wish. *logit.data* is a list of the components of the JAGS model we are

TABLE 6.1 JAGS code for Bayesian logistic model

```
X <- model.matrix(~ cdoc + female + kids + cage,
                    data = R84)
K <- ncol(X)
logit.data <- list(Y    = R84$outwork,
                   N    = nrow(R84),
                   X    = X,
                   K    = K,
                   LogN = log(nrow(R84)),
                   b0   = rep(0, K),
                   B0   = diag(0.00001, K)
                   )
sink("LOGIT.txt")

cat("
model{
    # Priors
    beta  ~ dmnorm(b0[], B0[,])

    # Likelihood
    for (i in 1:N){
      Y[i] ~ dbern(p[i])
      logit(p[i]) <- max(-20, min(20, eta[i]))
      eta[i]       <- inprod(beta[], X[i,])
      LLi[i] <- Y[i] * log(p[i]) +
                (1 - Y[i]) * log(1 - p[i])
  }
  LogL <- sum(LLi[1:N])
  AIC <- -2 * LogL + 2 * K
  BIC <- -2 * LogL + LogN * K
}
",fill = TRUE)
sink()

# INITIAL VALUES - BETAS AND SIGMAS
inits <- function () {
  list(
    beta  = rnorm(K, 0, 0.1)
    )  }
params <- c("beta", "LogL", "AIC", "BIC")

# JAGs
J0 <- jags(data = logit.data,
           inits = inits,
           parameters = params,
           model.file = "LOGIT.txt",
           n.thin = 10,
           n.chains = 3,
           n.burnin = 40000,
           n.iter   = 50000)

# OUTPUT DISPLAYED
out <- J0$BUGSoutput
myB <- MyBUGSOutput(out, c(uNames("beta", K), "LogL", "AIC", "BIC"))
round(myB, 4)
```

developing. *Y* is the response variable, *N* is the number of observations in the model, *X* is the predictor, *K* is the number of predictors, log *N* is the log of the number of observations in the model, b0 is the prior mean and B0 the prior precision, which is the inverse of the variance. b0 and B0 relate to defining priors. Arguments to b0 and B0 define the mean and variance for each prior in the model. In the code below, b0 is a vector of *K* priors, each with a value of 0; B0 indicates a matrix with a diagonal having *K* trace terms and a variance of each having values of 100,000.

```
logit.data <- list(Y    = R84$outwork,
                   N    = nrow(R84),
                   X    = X,
                   K    = K,
                   LogN = log(nrow(R84)),
                   b0   = rep(0, K),
                   B0   = diag(0.00001, K)
                   )
```

The next segment contains the terms

```
sink("LOGIT.txt")
cat("
```

that puts everything within the model braces, { }, below the code into a text file called *LOGIT.txt*.

Within the defining characteristics of the model, within the parentheses

```
model{
```

We start by defining the prior. The prior betas are defined as multivariate normal for all. The values we just defined for both b0 and B0 are supplied to the arguments of *dmnorm*().

```
beta ~ dmnorm(b0[], B0[,])
```

If we wanted to have a uniform prior for each of the predictors, the right side of the above distribution would be expressed as *dunif(-20, 20)*.

The following code segment defines the likelihood. This is a crucial segment. The likelihood is calculated across all observations in the model; that is, from 1 to *N*.

The first term in the for-loop specifies this to be a logistic regression; that is, the likelihood across all observations is Bernoulli distributed. The next two lines provide the link function, *logit* and *eta*, which is the linear predictor.

It is formed by the product (*inprod*) of the beta and *X* values. The final line within the parenthesis is the Bernoulli log-likelihood function. The sum of the observation log-likelihood values produces the model log-likelihood statistic, *LogL*.

```
   for (i in 1:N){
      Y[i] ~ dbern(p[i])
      logit(p[i]) <- max(-20, min(20, eta[i]))
      eta[i]       <- inprod(beta[], X[i,])
      LLi[i] <- Y[i] * log(p[i]) +
                (1 - Y[i]) * log(1 - p[i])
   }
   LogL <- sum(LLi[1:N])
```

The Akaike and Bayesian information criteria (AIC and BIC) statistics are then calculated, and the model parenthesis closes. The *fill* confirms that the LOGIT.txt file should contain everything within the parenthesis, and the *sink*() function actually saves the file to the working directory.

```
  AIC <- -2 * LogL + 2 * K
  BIC <- -2 * LogL + LogN * K
}
",fill = TRUE)
sink()
```

The *inits* segment formally defines the initial parameter values, which are all defined as normally distributed terms with a mean of 0 and variance of 10 (the precision, 1/*V*, is 0.1). The term *params* contains the coefficient, log-likelihood, AIC, and BIC statistics.

```
inits <- function () {
  list(
    beta   = rnorm(K, 0, 0.1)
    ) }

params <- c("beta", "LogL", "AIC", "BIC")
```

The segment JO is the JAGS function, containing the values and settings we just defined. The JAGS algorithm itself uses the following values to define the manner in which MCMC sampling occurs. It is the core of the JAGS function.

Terms we have not defined yet include the *n.thin*, meaning here that sampling actually keeps every 10th value from the MCMC Gibbs sampler, discarding the others. This is done in case the data are autocorrelated. Thinning is an attempt to increase sampling efficiency. Keeping one of every 10 samples for our distribution helps effect randomness. *n.chains* specifies how many distributions are being sampled. Chains of sampling are mixed, which assists in obtaining a distribution that properly characterizes the data. Here we specify that three chains of sampling are to be run. *n.burnin* indicates how many sampling values are discarded before values are kept in the posterior distribution. The initial values can range all over, the later skew the results. If all of the early values were kept, the mean of the posterior distribution could be severely biased. Discarding a sizeable number of early values helps guarantee a better posterior. Finally, the *n.iter* specifies how many values are kept for the posterior distribution, after thinning and discarding of burn-in values.

```
J0 <- jags(data = logit.data,
           inits = inits,
           parameters = params,
           model.file = "LOGIT.txt",
           n.thin = 10,
           n.chains = 3,
           n.burnin = 40000,
           n.iter   = 50000)
```

The last code in Table 6.1 is the output code. The source code in *jhbayes.R* is relevant at this point. The posterior means, or betas, the log-likelihood function, and AIC and BIC statistics are displayed, together with their standard errors and outer 0.025 "credible set." We specified that only four decimal digits are displayed. *BUGSoutput* and *MyBUGSOutput* are parts of the *R2jags* package:

```
out <- J0$BUGSoutput
myB <- MyBUGSOutput(out, c(uNames("beta", K),
      "LogL", "AIC", "BIC"))
round(myB, 4)
```

The Bayesian logistic model results are listed in the table below.

```
> round(myB, 4)
               mean     se        2.5%       97.5%
beta[1]     -2.0193 0.0824     -2.1760     -1.8609
beta[2]      0.0245 0.0063      0.0128      0.0370
beta[3]      2.2569 0.0843      2.0922      2.4216
beta[4]      0.3685 0.0904      0.1920      0.5415
beta[5]      0.0545 0.0042      0.0466      0.0626
LogL     -1961.6258 1.5178 -1965.4037 -1959.5816
AIC       3933.2517 3.0357  3929.1632  3940.8074
BIC       3964.5619 3.0357  3960.4734  3972.1176
```

Compare the above statistics with the summary table of *myg,* which was the model as estimated using the *glm* function. Note that the AIC values are statistically identical. This output also matches the SAS results displayed estimated using the noninformative prior.

```
> summary(myg)
Coefficients:
             Estimate Std. Error z value Pr(>|z|)
(Intercept) -2.010276   0.081087 -24.792  < 2e-16 ***
cdoc         0.024432   0.006263   3.901 9.57e-05 ***
female       2.256804   0.082760  27.269  < 2e-16 ***
kids         0.357976   0.089962   3.979 6.92e-05 ***
cage         0.054379   0.004159  13.075  < 2e-16 ***
---
    Null deviance: 5091.1  on 3873  degrees of freedom
Residual deviance: 3918.2  on 3869  degrees of freedom
AIC: 3928.2
```

A comparison of the frequency-based standard logistic regression and our two Bayesian models without informative priors produce nearly identical values. Note that using two entirely different methods of estimation—maximum likelihood and sampling—result in the same values. This tells us that these estimation procedures are valid ways of estimating the true underlying parameter values of the distribution theoretically generating the data.

```
> round(cbind(coef(myg), Bcoef, myB[1:K,1]), 4)
                      Bcoef
(Intercept) -2.0103 -2.0131 -2.0193
cdoc         0.0244  0.0246  0.0245
female       2.2568  2.2592  2.2569
kids         0.3580  0.3575  0.3685
cage         0.0544  0.0544  0.0545
```

The example above did not employ an informative prior. For instance, we could have provided information that reflected our knowledge that *docvis* has between 40% and 50% zero counts. We compounded the problem since *docvis* was centered, becoming *cdoc*. The centered values for when $docvis = 0$ are –3.162881. They are –2.162881 when $docvis = 1$. We can therefore set up a prior that we expect 40%–50% zero counts when *cdoc* is less than –3.

6.2.3 Bayesian Logistic Regression with Informative Priors

Priors placed on the predictor or parameter betas can be on the coefficients or slopes, or on the predictor data itself. The example of *cdoc* above was on the data. We could also set a prior on the coefficient of *cdoc* such that we are

75% confident that the coefficient will be between 0.020 and 0.030. The prior is expressed in terms of probability functions, usually the normal, lognormal, beta, binomial, Bernoulli, Cauchy, *t*, gamma, inverse gamma, Poisson, Poisson gamma, and negative binomial. When one prior is specified on the prior line, it is assumed to be applied to all model predictors. Otherwise, each predictor, including the intercept, may have a prior.

The example below employs Cauchy prior on all three parameters; that is, the *intercept*, *cdoc*, and *cage*.

beta.0 ~ d*t*(0,1/(2.5^2),1)
beta.1 ~ d*t*(0, 1/(2.5^2),1)
beta.2 ~ d*t*(0, 1/(2.5^2),1)

where 1/(2.5)^2 is equal to 0.16. For those of my readers who have taken a course in probability, you may recall that the Cauchy corresponds to a Student's *t* distribution, with $2n - 1$ degrees of freedom, multiplied by the value $1/\text{sqrt}(s*(2*n - 1))$. *n* and *s* are the shape and scale parameters, respectively, for the Cauchy distribution. Perhaps the normal might be preferable for the intercept, and an analyst should check to see if it is the case (Table 6.2). The code, presented in a slightly different manner to Table 6.1 can be used for a wide variety of models. The output does not include implementing the *R2jags MyBUGSOutput* function that produces nicely formatted results.

```
#  load contents of Table 6.2 into memory prior to
   running summary() below
>  summary(codasamples)

Iterations = 41001:91000
Thinning interval = 1
Number of chains = 3       # <= note that 3 chains are used
Sample size per chain = 50000

1. Empirical mean and standard deviation for each variable,
   plus standard error of the mean:

              Mean        SD  Naive SE Time-series SE
AIC      5.020e+03 1.996398 0.0051547      7.634e-03
BIC      5.026e+03 1.996398 0.0051547      7.634e-03
LogL    -2.509e+03 0.998199 0.0025773      3.817e-03
beta.0 -5.481e-01 0.047049 0.0001215      8.214e-04
beta.1  5.059e-02 0.006352 0.0000164      2.072e-05
beta.2  7.209e-02 7.991224 0.0206333      2.006e-01

2. Quantiles for each variable:

             2.5%        25%        50%        75%      97.5%
AIC     5.018e+03  5.018e+03  5.019e+03  5.020e+03  5.025e+03
```

TABLE 6.2 JAGS logistic regression with cauchy prior

```
library(R2jags)
library(COUNT)
data(rwm1984)
R84 <- rwm1984
R84$cage <- R84$age - mean(R84$age)
R84$cdoc <- R84$docvis - mean(R84$docvis)

### JAGS component
K <- 1
logit.data <- list(Y    = R84$outwork,
                   N    = nrow(R84),
                   cdoc = R84$cdoc,
                   cage = R84$cage,
                   K=1,
                   LogN = log(nrow(R84))
)
GLM.txt<-"
    model{
    #1. Priors
    beta.0~ dt(0,.16, 1)
    beta.1~ dt(0, .16, 1)
    beta.2~ dt(0, .16,1)

    #2. Likelihood
    for (i in 1:N){

    Y[i] ~ dbern(p[i])
    logit(p[i]) <- max(-20, min(20, eta[i]))
    eta[i]      <- beta.0+beta.1*cdoc[i]+beta.2*cage[2]

    LLi[i] <- Y[i] * log(p[i]) +
    (1 - Y[i]) * log(1 - p[i])
    }
    LogL <- sum(LLi[1:N])
    AIC <- -2 * LogL + 2 * K
    BIC <- -2 * LogL + LogN * K

    }
    "
# INITIAL VALUES - BETAS AND SIGMAS
inits <- function () {
  list(
    beta.0  = 0.1,beta.1=0.1, beta.2=0.1
  ) }
params <- c("beta.0","beta.1","beta.2","LogL", "AIC", "BIC")

# JAGs
J0 <- jags.model(data = logit.data,
           inits = inits,
           textConnection(GLM.txt),
           n.chains = 3,
           n.adapt=1000)
update(J0, 40000)
codasamples <- coda.samples(J0, params, n.iter = 50000)
summary(codasamples)
```

```
BIC      5.024e+03  5.024e+03  5.025e+03  5.027e+03  5.031e+03
LogL    -2.511e+03 -2.509e+03 -2.508e+03 -2.508e+03 -2.508e+03
beta.0 -6.282e-01 -5.725e-01 -5.476e-01 -5.230e-01 -4.682e-01
beta.1  3.839e-02  4.629e-02  5.050e-02  5.481e-02  6.325e-02
beta.2 -1.112e+01 -9.996e-01 -7.193e-03  9.644e-01  1.070e+01
```

With normal prior, the output is displayed as:

```
#1. Priors
beta.0 ~ dnorm(0, 0.00001)
beta.1~dnorm(0, 0.00001)
beta.2~dnorm(0, 0.00001)
```

```
1. Empirical mean and standard deviation for each variable,
   plus standard error of the mean:

             Mean       SD  Naive SE Time-series SE
AIC     5.020e+03 1.983282 5.121e-03      7.678e-03
BIC     5.026e+03 1.983282 5.121e-03      7.678e-03
LogL   -2.509e+03 0.991641 2.560e-03      3.839e-03
beta.0 -5.471e-01 0.044242 1.142e-04      5.766e-04
beta.1  5.058e-02 0.006379 1.647e-05      2.113e-05
beta.2 -1.546e-01 7.041635 1.818e-02      1.412e-01

2. Quantiles for each variable:

             2.5%        25%        50%        75%      97.5%
AIC      5.018e+03  5.018e+03  5.019e+03  5.020e+03  5.025e+03
BIC      5.024e+03  5.024e+03  5.025e+03  5.027e+03  5.031e+03
LogL    -2.511e+03 -2.509e+03 -2.508e+03 -2.508e+03 -2.508e+03
beta.0 -6.255e-01 -5.722e-01 -5.473e-01 -5.223e-01 -4.660e-01
beta.1  3.829e-02  4.625e-02  5.051e-02  5.484e-02  6.333e-02
beta.2 -1.235e+01 -1.022e+00 -2.184e-02  9.511e-01  9.943e+00
```

Notice that the values of the distributional means for each parameter—*intercept*, *cdoc*, and *cage*—differ, as do other associated statistics. The prior has indeed changed the model. What this means is that we can provide our model with a substantial amount of additional information about the predictors used in our logistic model. Generally speaking, it is advisable to have a prior that is distributionally compatible with the distribution of the predictor having the prior. The subject is central to Bayesian modeling, but it takes us beyond the level of this book. My recommendations for taking the next step in Bayesian modeling include Zuur et al. (2013), Cowles (2013), and Lunn et al. (2013). More advanced but thorough texts are Christensen et al. (2011) and Gelman et al. (2014). There are many other excellent texts as well. I should also mention that Hilbe et al. (2016) will provide a clear analysis of Bayesian modeling as applied to astronomical data.

SAS CODE

```
/* Section 6.2 */
*Refer to the code in section 1.4 to import and print rwm1984 dataset;
*Refer to proc freq in section 2.4 to generate the frequency table;
*Summary for continuous variables;
proc means data=rwm1984 min q1 median mean q3 max maxdec=3;
        var docvis age;
        output out=center mean=;
run;


*Create the macro variables;
proc sql;
        select age into: meanage from center;
        select docvis into: meandoc from center;
quit;

*Center the continuous variables;
data R84;
        set rwm1984;
        cage=age-&meanage;
        cdoc=docvis-&meandoc;
run;


*Build the logistic model and obtain odds ratio & statistics;
proc genmod data=R84 descending;
        model outwork=cdoc female kids cage / dist=binomial link=logit;
        estimate "Intercept" Intercept 1 / exp;
        estimate "Cdoc" cdoc 1 / exp;
        estimate "Female" female 1 / exp;
        estimate "Kids" kids 1 / exp;
        estimate "Cage" cage 1 / exp;
run;


*Build the quasibinomial logistic model;
proc glimmix data=R84;
        model outwork (event='1')=cdoc female kids cage / dist=binary
        link=logit solution covb;
        random _RESIDUAL_;
run;

*Refer to proc iml in section 2.3 and the full code is provided
online;
```

```
*Bayesian logistic regression;
proc genmod data=R84 descending;
        model outwork=cdoc female kids cage / dist=binomial link=logit;
        bayes seed=10231995 nbi=5000 nmc=100000
        coeffprior=uniform diagnostics=all
        statistics=(summary interval) plots=all;
run;

*Create the normal prior;
data prior;
        input _type_ $ Intercept cdoc cage;
        datalines;
Var 1e5 1e5 1e5
Mean 0 0 0
;
run;

*Bayesian logistic regression with normal prior;
proc genmod  data=R84 descending;
        model outwork=cdoc female kids cage/dist=binomial link=logit;
        bayes seed=10231995 nbi=5000 nmc=100000
        coeffprior=normal(input=prior) diagnostics=all
        statistics=(summary interval) plots=all ;
run;
```

SAS output Bayesian logistic regression with normal prior.

INDEPENDENT NORMAL PRIOR FOR REGRESSION COEFFICIENTS

PARAMETER	*MEAN*	*PRECISION*
Intercept	0	0.00001
Cdoc	0	0.00001
Female	0	1E-6
Kids	0	1E-6
Cage	0	0.00001

STATA CODE

```
. use rwm1984
. center docvis, pre(c)
. rename cdocvis cdoc
. center age, pre(c)
. sum cdoc cage
* Logistic regression: standard and scaled
. glm outwork cdoc female kids cage, fam(bin) eform nolog
. glm outwork cdoc female kids cage, fam(bin) eform scale(x2) nolog
* Non-informative priors, normal(0, 100000)
```

```
. bayesmh outwork cdoc female kids cage, likelihood(logit) prior({outwork:},normal(
   0, 100000))
. bayesgraph diagnostics {outwork:}
. bayesstats ic
* Informative priors: Cauchy prior on cdoc and cage; noninformative on others
. bayesmh outwork cdoc female kids cage, likelihood(logit)            ///
       prior({outwork:female kids _cons}, normal(0, 100000))          ///
       prior({outwork:cdoc},                                          ///
       logdensity(ln(6.276)-ln(6.276^2+({outwork_cdoc})^2)-ln(_pi))) ///
       prior({outwork:cage},                                          ///
       logdensity(ln(11.24)-ln(11.24^2+({outwork_cage})^2)-ln(_pi))) ///
       block({outwork:female kids _cons})
. bayesgraph diagnostics {outwork: cdoc}
. bayesgraph diagnostics {outwork; cage)
. bayesstats ic
```

Stata 14: Partial Output—Logit Model with Informative Priors

```
Bayesian logistic regression                  MCMC iterations  =     12,500
Random-walk Metropolis-Hastings sampling      Burn-in          =      2,500
                                              MCMC sample size =     10,000
                                              Number of obs    =      3,874
                                              Acceptance rate  =      .1792
                                              Efficiency:  min =     .05461
                                                           avg =     .07162
Log marginal likelihood = -27363.562                       max =     .09621
```

outwork	Mean	Std. Dev.	MCSE	Median	Equal-tailed [95% Cred.	Interval]
cdoc	.0199813	.0055174	.000178	.020052	.0091184	.029969
female	2.243112	.0815326	.003024	2.241708	2.092979	2.403369
kids	.2936242	.0886074	.003792	.2910353	.1282263	.4675729
cage	.0485316	.0040249	.000149	.048308	.0408274	.0566746
_cons	-1.967223	.0791867	.003202	-1.970386	-2.131315	-1.803809

CONCLUDING COMMENTS

This book is intended as a guidebook to help analysts develop and execute well-fitted logistic models. In reviewing it now that it is finished, the book can also be regarded as an excellent way for an analyst to learn R, as well as SAS and Stata as applied to developing logistic models and associated tests and data management tasks related to statistical modeling. Several new functions are found in this book that are new to R—functions that were written to assist the analyst in producing and testing logistic models. I will frequently use these functions in my own future logistic modeling endeavors.

I mentioned in the book that when copying code from one electronic format to another, characters such as quotation marks and minus signs can result in errors. Even copying code from my own saved Word and PDF documents to R's editor caused problems. Many times I had to retype quotation marks, minus signs, and several other symbols in order for R to run properly. I also should advise you that when in the R editor, it may be wise to "run" long stretches of code in segments. That is, rather than select the entire program code, select and run segments of it. I have had students, and those who have purchased books of mine that include R code, email me that they cannot run the code. I advise them to run it in segments. Nearly always they email back that they now have no problems. Of course, at times in the past there have indeed been errors in the code, but know that the code in this book has all been successfully run multiple times. Make sure that the proper libraries and data have been installed and loaded before executing code.

There is a lot of information in the book. However, I did not discuss issues such as missing values, survey analysis, validation, endogeny, and latent class models. These are left for my comprehensive book titled, *Logistic Regression Models* (2009, Chapman & Hall), which is over 650 pages in length. A forthcoming second edition will include both Stata and R code in the text with SAS code as it is with this book. Bayesian logistic regression will be more thoroughly examined, with Bayesian analysis of grouped, ordered, multinomial, hierarchical, and other related models addressed.

I primarily wrote this book to go with a month-long web-based course I teach with Statistics.com. I have taught the course with them since 2003, three classes a year, and continually get questions and feedback from researchers, analysts, and professors from around the world. I have also taught logistic regression and given workshops on it for over a quarter a century. In this book, I have tried to address the most frequent concerns and problem areas that practicing analysts have informed me about. I feel confident that anyone reading carefully through this relatively brief monograph will come away from it with a solid knowledge of how to use logistic regression—both observation based and grouped. For those who wish to learn more after going through this book, I recommend my *Logistic Regression Models* (2009, 2016 in preparation). I also recommend Bilger and Loughin (2015), which uses R code for examples, Collett (2003), Dohoo et al. (2012), and for nicely written shorter books dealing with the logistic regression and GLM in general, Dobson and Barnett (2008), Hardin and Hilbe (2013), and Smithson and Merkle (2014). Hosmer et al. (2013) is also a fine reference book on the subject, but there is no code provided with the book. The other recommended books have code to support examples, which I very much believe assists the learning process.

I invite readers of this book to email me their comments and suggestions about it: hilbe//works.bepress.com/joseph_hilbe/, has the data sets used in the book in various formats, and all of the code used in the book in electronic format. Both SAS and Stata code and output is also provided.

References

Bilder, C.R. and Loughin, T.M. 2015. *Analysis of Categorical Data with R*. Boca Raton, FL: Chapman & Hall/CRC.
Christensen, R., Johnson, W., Branscu, A. and Hanson, T.E. 2011. *Bayesian Ideas and Data Analysis*. Boca Raton, FL: Chapman & Hall/CRC.
Collett, D. 2003. *Modeling Binary Data*, 2nd Edn. Boca Raton, FL: Chapman & Hall/CRC.
Cowles. M.K. 2013. *Applied Bayesian Statistics*. New York, NY: Springer.
De Souza, R.S. Cameron, E., Killedar, M., Hilbe, J., Vilatia, R., Maio, U., Biffi, V., Riggs, J.D. and Ciardi, B., for the COIN Collaboration. 2015. The overlooked potential of generalized linear models in astronomy—I: Binomial regression and numerical simulations, *Astronomy & Computing*, DOI: 10.1016/j.ascom.2015.04.002.
Dobson, A.J. and Barnett, A.G. 2008. *An Introduction to Generalized Linear Models*, 3rd Edn. Boca Raton, FL: Chapman & Hall/CRC.
Dohoo, I., Martin, W. and Stryhn, H. 2012. *Methods in Epidemiological Research*. Charlottetown, PEI, CA: VER.
Firth, D. 1993. Bias reduction of maximum likelihood estimates, *Biometrika* 80, 27–28.
Gelman, A., Carlin, J.B., Stern, H.S., Dunson, D.B., Vehtari, A. and Rubin, C.B. 2014. *Bayesian Data Analysis,* 3rd Edn. Boca Raton, FL: Chapman & Hall/CRC.
Geweke, J. 1992. Evaluating the accuracy of sampling-based approaches to calculating posterior moments. In Bernardo, J.M., Berger, J.O., Dawid, A.P., Smith, A.F.M. (eds.), Bayesian Statistics, 4th Edn. Oxford, UK: Clarendon Press.
Hardin, J.W. and Hilbe, J.M. 2007. *Generalized Linear Models and Extensions*, 2nd edition, College Station, TX: Stata Press.
Hardin, J.W. and Hilbe, J.M. 2013. *Generalized Linear Models and Extensions*, 3rd Edn., College Station, TX: Stata Press/CRC (4th edition due in 2015).
Hardin, J. W. and Hilbe, J.M. 2014. Estimation and testing of binomial and beta-binomial regression models with and without zero inflation, *Stata Journal* 14(2): 292–303.
Heinze, G. and Schemper, M. 2002. A solution to the problem of separation in logistic regression. *Statistics in Medicine* 21, 2409–2419.
Hilbe, J.M. 2009. *Logistic Regression Models*. Boca Raton, FL: Chapman & Hall/CRC.
Hilbe, J.M. 2011. *Negative Binomial Regression*, 2nd Ed. Cambridge, UK: Cambridge University Press.
Hilbe, J.M. 2014. *Modeling Count Data*. New York, NY: Cambridge University Press.
Hilbe, J.M. and Robinson, A.P. 2013. *Methods of Statistical Model Estimation*. Boca Raton, FL: Chapman & Hall/CRC.
Hilbe, J.M., de Souza, R.S. and Ishida, E. 2016. *Bayesian Analysis of Astrophysical Data: Using R/JAGS and Python/Stan*. Cambridge, UK: Cambridge University Press.
Hosmer, D.W., Lemeshow, S. and Sturdivant, R.X. 2013. *Applied Logistic Regression*, 3rd Edn. Hokoken, NJ: Wiley.

Lunn, D., Jackson, C., Best, N., Thomas, A. and Speigelhalter, D. 2013. *The BUGS Book.* Boca Raton, FL: Chapman & Hall/CRC.

McGrayne, S.B. 2011. *The Theory that Would not Die.* New Haven, CT: Yale University Press.

Morel, G. and Neerchal, N.K. 2012. *Overdispersion Models in SAS.* Carey, NC: SAS Publishing.

Rigby, R.A. and Stasinopoulos, D.M. 2005. Generalized additive models for location, scale and shape, (with discussion). *JRSS Applied Statistics* 54: 507–554.

Smithson, M. and Merkle, E.C. 2014. *Generalized Linear Models for Categorical and Continuous Limited Dependent Variables.* Boca Raton, FL: Chapman & Hall/CRC.

Weisberg, H.I. 2014. *Willful Ignorance.* Hoboken, NJ: Wiley.

Youden, W.J. 1950. Index for rating diagnostic tests. *Cancer* 3: 32–35.

Zuur, A.F. 2012. *A Beginner's Guide to Generalized Additive Models with R.* Newburgh, UK: Highlands Statistics.

Zuur, A.F., Hilbe, J.M. and Ieno, E.M. 2013. *A Beginner's Guide to GLM and GLMM with R: A Frequentist and Bayesian Perspective of Ecologists.* Newburgh, UK: Highlands Statistics.

Index

A

Akaike information criterion (AIC), 72, 141; *see also* Bayesian information criterion (BIC)
  AIC_H statistic, 60
  statistics, 110
  test, 58–59
Apparent overdispersion, 114, 115
Area under curve (AUC), 85, 86

B

badhealth data, 34, 35, 36
Basic model statistics, 20; *see also* Logistic models
  confidence intervals, 24–28
  p-value, 23–24
  standard errors, 20–23
  z statistics, 23
Bayesian inference Using Gibbs Sampling (BUGS), 129, 130, 138
Bayesian information criterion (BIC), 60; *see also* Akaike information criterion (AIC)
  statistics, 141, 142
Bayesian logistic regression, 129–130; *see also* Grouped logistic regression
  informative priors, 143–147
  JAGS, 137–143
  R, 130–137
  R trace and density plots of model, 135
Bayesian methodology, 127
  basic formula, 128
  Bayesian logistic regression, 129
  characteristic features, 128
Bernoulli distribution, 2, 4
  likelihood function, 5
  linear predictor of logistic model, 7
  log-likelihood, 53
  probability function, 4, 5
  sum of regression terms, 6
Bernoulli logistic regression, residuals for, 74
Beta-binomial regression, 117; *see also* Grouped logistic regression
  beta-binomial model, 123
  beta binomial, 119, 122
  beta PDF, 118
  binomial PDF, 117
  covariate pattern, 120
  log-likelihood function for binomial model, 118
  odds ratio for beta binomial, 122
  robust adjustments of standard errors, 121
Beta binomial, 114, 119, 122, 123, 144
  distribution, 119
  model, 123
  modeling binomial component of, 117
  odds ratio for, 122
  R output for, 121
BIC, *see* Bayesian information criterion (BIC)
Binary predictor, 13; *see also* Continuous predictor; Categorical predictor
  model-based confidence intervals, 15
  odds-intercept, 17
  odds ratio, 16, 17
  R's *glm* function, 14
  summary function, 14
Binary variable, 3, 28, 65, 75, 98, 99
Binomial probability distribution function, 107–108
Bootstrapping, 64–65
BUGS, *see* Bayesian inference Using Gibbs Sampling (BUGS)

C

CABG, *see* Coronary artery bypass grafting (CABG)
Categorical predictor, 28; *see also* Binary predictor; Continuous predictor
  emergency level of type, 31–32
  factor function, 30
  internal indicator variables, 30

Categorical predictor (*Continued*)
interpretation, 30, 31
no-frills frequency table, 29
provnum string variable, 28
centered *age* (*cage*), 36, 136
centered *docvis* (*cdoc*), 136
Centering, 34
badhealth data, 34, 35
continuous predictors, 131
numvisit, 35
scale function, 36
Classification statistics, 81
confusion matrix, 86–88
medpar data, 82
ROC analysis, 84–86
S–S plot, 83, 84
Conditional effects plot, 79, 81
code for creating, 80
graph, 103
preparation, 102–103
variable *type*, 80
Confidence intervals, 10, 11, 20, 26, 128; *see also* Profile confidence intervals
confint.default(), 27
glm function, 25
medpar data, 26
model-based, 15, 20
odds ratios, 24
qnorm function, 24
toOR function, 25
confint() function, 11, 20, 27
confint.default() function, 11, 20, 27
Confounders, 65–67
Confusion matrix, 83, 86
classification, 88
classification analysis, 86
PresenceAbsence package, 87
ROC, 83
Conjugate relationship, 118
Continuous predictor, 32, 34, 77; *see also* Binary predictor; Categorical predictor
centering, 34–36
curves, 80
GAM, 33–34
interpretation, 33
"linear in logit", 32
log transform, 33
standardization, 36–37
Coronary artery bypass grafting (CABG), 94
Covariate patterns, 4, 76, 91, 115, 120

D

delta method, 22, 23
Dependent variable, *see* Response variable
Deviance residuals, 14, 56
squared standardized, 78, 102
standardized, 57, 78
statistics, 42
Deviance statistic, 53, 108
deviance residuals, 56
logistic model, 54
drop1 function, 72, 73

E

edf, *see* effective degrees of freedom (edf)
educlevel predictor, 50, 51, 52, 54
effective degrees of freedom (edf), 34
Effect modifiers, 65–67
Estimation methods, 7
confint function, 11
convergence, 136
glm function, 8
logistic regression, 25
MLE, 7
mylogit, 10
R function for logistic regression, 9
Exact logistic regression, 93; *see also* Bayesian logistic regression; Grouped logistic regression
CABG and PTCA, 94
p-value of *procedure*, 95–96
quasibinomial model, 95
Exponential family form, 5, 53, 107
Exponentiation, 4, 98
coefficient-based confidence intervals, 24
intercept, 16
logged value, 5

F

factor function, 30
False negatives, 83
False positive, 24, 83, 84
Finite sample, 59

G

GAM, *see* Generalized additive model (GAM)
gamlss function, 119, 121, 123

Gaussian distribution, 2, 6
Generalized additive model (GAM), 33–34
  los, 35
  partial residual plots and use, 33
Generalized linear model (GLM), 8
  binomial distribution, 117
  framework, 8
  glm function, 8, 14, 22, 117
  logistic regression, 52, 150
  statistics, 53
GLM, *see* Generalized linear model (GLM)
Goodness-of-fit (GOF), 53, 71
  logistic model Pearson Chi2 GOF statistic, 55
  Pearson *Chi*2, 54, 71–72
  test, 54
Grouped data, 97, 110
  observation to, 109–112
  response variable, 110
  Stata and SAS, 111
Grouped logistic regression; *see also* Bayesian logistic regression; Beta-binomial regression
  binomial probability distribution function, 107–108
  identifying and adjusting for extra dispersion, 113–115
  modeling and interpretation, 115–116
  observation to grouped data, 109–112

H

Hat matrix diagonal, 57
  influence statistic, 57
Highest posterior density region (HPD region), 128
hiv data, 91, 92
Hosmer–Lemeshow test (H–L test), 88
  H–L *Chi*2 test, 88, 89
  H–L tables, 90–91
  medpar data, 88
HPD region, *see* Highest posterior density region (HPD region)

I

ibeta function, 73, 74
Implicit overdispersion correlation, 113
Information criterion tests, 58
  AIC test, 58–59
  BIC, 60
  finite sample, 59
Informative priors
  Bayesian logistic regression, 143
  Bayesian modeling, 146–147
  Cauchy prior, 144
  JAGS logistic regression with Cauchy prior, 145–146
  partial output—logit model with, 149
Interactions, 65–67
Internal indicator variables, 30
Inverse link function, 108
Inverse probability, 127, 128
IRLS algorithm, *see* Iterative reweighted least squares algorithm (IRLS algorithm)
Iterative reweighted least squares algorithm (IRLS algorithm), 8, 93

J

Just Another Gibbs Sampler (JAGS), 137
  AIC and BIC, 141
  Bayesian logistic model, 142
  Bayesian logistic regression, 137
  code for Bayesian logistic model, 139
  frequency-based standard logistic regression, 143
  likelihood, 140
  logistic regression with Cauchy prior, 145–146
  R's path, 138

L

Length of Stay (LOS), 33–34, 80
  GAM model, 35
  probability, 41
Likelihood function, 5, 129
Likelihood ratio test, 27, 72–73, 93
"Linear in logit", 32
Linear predictor, 6, 15, 49, 140
  fitted values, 18, 33
  logistic model, 7
  standard error, 39
Link function, 6, 7, 15, 108, 140
  inverse logistic, 18, 40
  logistic, 32
  true inverse logit, 40
linkinv function, 40

Log-likelihood function, 6, 56–57, 108, 142
  Bernoulli, 5, 141
  binomial model, 118
  deviance statistic, 53
log-odds, 15, 25, 56, 83
logistf() function, 92–93
Logistic model, 13, 49; *see also* Statistical model
  Bernoulli deviance, 54
  Bernoulli distribution log-likelihood, 53
  binary predictor, 13–17
  categorical predictor, 28–32
  conditional effects plot, 79–81
  confounders, 65–67
  continuous predictor, 32–37
  effect modifiers, 65–67
  information criterion tests, 58–60
  interactions, 65–67
  likelihood ratio test, 72–73
  model fitting process, 61–65
  odds ratios, 18–20
  Pearson *Chi*2 goodness-of-fit test, 71–72
  predictions, 18–20, 37–41
  probabilities, 18–20
  R *summary* function, 52–53
  residual analysis, 73–79
  risk factors, 65–67
  SAS code, 41–47, 67–70
  selection and interpretation of predictors, 49–52
  STATA code, 47–48, 70
  statistics in, 52–57
Logistic model deviance statistic, 54
Logistic Model Pearson Chi2 GOF Statistic, 55
Logistic regression, 18, 50, 81, 150; *see also* Bayesian logistic regression; Exact logistic regression; Grouped logistic regression
  Bernoulli-based, 6
  Bernoulli variance function, 54
  classification tools, 81
  glm, 25
  logic of modeling data with, 49
  modeling, 3–4
  parameter estimates, 58
  penalized, 92, 93
  residuals for Bernoulli, 74
  R function for, 9
  R's default, 8
  statistical software for, 20
logit, *see* log-odds
logit1 model name, 13
Log transform, 33
LOS, *see* Length of Stay (LOS)

M

m-asymptotic form, 75, 77
Markov Chain Monte Carlo model (MCMC model), 132, 133
Maximum likelihood estimation (MLE), 7, 8
MCMC model, *see* Markov Chain Monte Carlo model (MCMC model)
Measurement error, 2
medpar data set, 8, 26, 33, 58, 77, 88
MLE, *see* Maximum likelihood estimation (MLE)
Model-based confidence intervals, 15, 20, 28
modelfit function, 59, 60
Model fitting process, 61
  bootstrapping, 64–65
  robust standard errors, 63–64
  scaling standard errors, 61–63
Model sensitivity, 82, 83, 84
Model specificity, 82
mu parameter, 6, 82, 116, 121
mylgg model, 115
mylogit model, 10
mymod model, 56, 60, 61, 72, 73

N

n-asymptotic form, 74, 77
Negative inverse Hessian matrix, 21
"New Script" editor, 8, 80
Null deviance, 54
numvisit model, 35

O

Observation data, 109, 112, 114
Odds-intercept, 17, 19, 20, 24
Odds ratio, 16, 17, 18–20, 33, 51
  beta binomial, 122
  confidence intervals for, 24
  predictors, 4
  standard errors of, 22
  z statistic for, 23

P

p-value, 10, 23, 63, 88, 122
  logistic model, 23
  model coefficients and, 93
  procedure, 95
  regression software, 24
Parameter(s), 2
  distribution, 5
  estimates, 7, 8, 20, 128
Parametric model, 1
Parametric statistical models, 1
PDF, *see* Probability distribution function (PDF)
Pearson *Chi*2 goodness-of-fit
  statistic, 54
  test, 71–72
Pearson dispersion statistic, 22, 62, 95, 114
Pearson residual, 55
  leverage *vs.* standardized, 115
  standardize, 57
Penalized logistic regression, 92, 93
Percutaneous transluminal coronary angioplasty (PTCA), 94
Poisson model, 113, 114
Population data, 2, 5
Prediction, 2, 18–20, 37
  basics, 37–39
  confidence intervals, 39–40
  los, 38
  probability of length of stay, 41
Predictors, 3, 27, 75, 91, 146
  binary, 109
  categorical, 29
  continuous, 32–33
  educlevel predictor, 50
  interpretation, 52
  matrix, 138
  odds ratios, 51
  selection and interpretation, 49
PresenceAbsence package, 87
Prior distribution, 128, 129
Probabilities, 4, 18–20
  model linear predictor and, 19
  predicted, 82
Probability distribution function (PDF), 1–2, 136
  Bernoulli, 61
  beta, 118, 119
  binomial, 107, 117
Probability function, 4, 5, 113, 144
"Problem of separation", 92
Profile confidence intervals, 11, 20, 27, 28
provnum string variable, 28
PTCA, *see* Percutaneous transluminal coronary angioplasty (PTCA)

Q

qnorm function, 24
Quasibinomial models, 22, 55, 95, 131
Quasipoisson option, 22

R

Receiver operator characteristic curve (ROC curve), 81, 85
  analysis, 84
  AUC statistic, 86
  ROCtest function, 85
Regression, 2; *see also* Bayesian logistic regression; Beta-binomial regression; Exact logistic regression; Grouped logistic regression
"Residual" deviance, 54
Residual analysis, 73
  Bernoulli logistic regression, 74
  leverage plot, 79
  m-asymptotic data, 76, 77
  residual code, 75
  squared standardized deviance *vs.* mu, 77, 78
Response term, 2, 18, 65, 66, 85
Response variable, 3, 4, 27, 82, 98, 140
Risk factors, 65–67
Robust standard errors, 63–64
ROC curve, *see* Receiver operator characteristic curve (ROC curve)
R *summary* function, 52–53

S

Sandwich standard errors, *see* Robust standard errors
SAS code
  Bayesian logistic regression, 147–148
  grouped logistic regression, 123–125
  logistic models, 41–47, 67–70, 101–105
  statistical model, 11–12
Scaled logistic model, 131

Scaling, 22
  p-value of *procedure*, 95
  standard errors, 61–63
Sensitivity–specificity plot (S–S plot), 81, 84, 88
Sigma, 121, 122
Specificity, 82–83, 84, 87
S–S plot, *see* Sensitivity–specificity plot (S–S plot)
Standard errors, 10, 20, 132
  bootstrapped, 64
  coefficient, 20
  negative inverse Hessian matrix, 21
  odds ratios, 22
  robust or sandwich, 63
  scaling, 61–63
Standardization, 36–37, 57
Standardized deviance residual, 57, 78, 115
Standardize Pearson residual, 57
STATA code
  Bayesian logistic regression, 148–149
  grouped logistic regression, 125–126
  logistic models, 47–48, 70, 105–106
  statistical model, 12
Statistical model, 1; *see also* Logistic models
  Bernoulli distribution, 4–7
  estimation methods, 7–11
  logistic regression modeling, 3–4
  parameters, 2
  parametric statistical models, 1
  SAS code, 11–12
  STATA code, 12
  "violations of distributional assumptions", 3
summary function, 14, 15, 20
Sum of regression terms, 6

T

Table data modeling, 96
  binary response, 99
  complete logistic model, 100
  data in grouped format, 98
  observation-level format, 97
titanicgrp data, 119
toOR function, 25
Two-parameter model, 114, 117

U

Unbalanced data, models with
  logistf() function, 92–93
  perfect prediction, 91
  "problem of separation", 92

V

vcov function, 21
Violations of distributional assumptions, 3

W

Wald confidence intervals, *see* Confidence intervals

Z

z statistics, 20, 23, 24